IEE TELECOMMUNICATIONS SERIES 50

Series Editors: Professor C. J. Hughes
Professor J. O'Reilly
Professor G. White

Telecommunications Regulation

Other volumes in this series:

Telecommunications Regulation

John Buckley

The Institution of Electrical Engineers

Published by: The Institution of Electrical Engineers, London,
United Kingdom

The Institution of Electrical Engineers,
Michael Faraday House,
Six Hills Way, Stevenage,
Herts., SG1 2AY, United Kingdom

British Library Cataloguing in Publication Data

Buckley, J.
    Telecommunications regulation. – (IEE telecommunications series ; 50)
    1. Telecommunication – Law and legislation
    2. Telecommunication – Management
    I. Title II. Institution of Electrical Engineers
    343'.0994

ISBN 0 85296 444 7

Typeset in India by Newgen Imaging Systems
Printed in the UK by MPG Books Limited, Bodmin, Cornwall

Contents

Foreword

The subject of regulation stimulates strong opinions. Almost nobody in the telecommunications industry can claim to be uninterested in the topic and indeed few are disinterested in its impact. In short, regulation matters.

But despite that, it is a subject that is widely misunderstood. Many members of the public look at the regulator as a 'watchdog' whose duty it is to right every wrong and keep order in the industry. Some go further and expect the regulator to act as a 'fairy godmother' who will magically deliver their every wish. Even players in the industry itself sometimes like to think that it is the regulator's job to ensure their business plan is viable and remains so whatever happens to market conditions. Some even think we (the regulators) can prevent companies from going bust. If only it were so. But then, how damaging it might be if we truly had the powers of the Sorcerer's Apprentice!

The reality is that all regulators are 'creatures of statute', whose role and powers are closely defined by law, albeit that their wider objectives are general enough to leave plenty of scope to adopt different tactics, as can be seen by the different approaches taken across the globe or even within the common framework in Europe.

As an engineer myself, I like to think of regulation as a form of negative feedback. It is a form of control mechanism that has the effect of limiting the worst extremes of market swing and instability, but is not so constraining as to prevent any given swing from occurring at all. It does not define how the market will be, but has a strong influence on how it behaves. It is there to mitigate those aspects of 'market failure' where intervention has a net beneficial effect. And that means that doing nothing is sometimes the right answer.

Inevitably, for such an important subject, much has been written about regulation in recent years. This includes articles in the tabloid newspapers, which complain about the 'scandal' of high prices or poor service: 'Something must be done', they proclaim. At the other end of the spectrum, some heavyweight books, mostly targeted at professional people in these disciplines, have been written about the economic and legal principles that underpin regulation.

The present book is an interesting new contribution to the literature. John Buckley attempts to cover the whole subject at a reasonable level of detail. This will give professionals and managers from various disciplines valuable insights into the concerns of both their own and other specialisms. The book includes an in-depth look at issues such as interconnection, numbering and other technical aspects of regulation that will be of interest to engineers working in the industry. Such engineers will find the book of great value in not skimming over the parallel professional issues studied by economists, lawyers and those interested in competition theory. Taken as a whole, it enables the reader to understand the subject and participate in the debates in a rounded way.

Where those debates take us in the long run is as yet unknown and John Buckley avoids trying to judge the success or otherwise of the various regulatory approaches around the world. This is wise, as what happens in the market is not the sole responsibility of the regulator and there is no opportunity for a 'control experiment' against which to judge the merits or otherwise of regulation. The best we can do is to observe the different outcomes around the world, although even this raises the objection that too many of the 'key parameters', as well as regulation itself, vary from one country to another.

What we, as regulators, do is under the constant gaze of the industry and the media. No doubt the transition taking place in the UK, at the time of writing, to create a single regulator, Ofcom, from five existing telecoms and media regulators will be watched with intense interest. Sometimes I can imagine that if Isaac Newton were alive today, he might rewrite his laws of motion around the telecommunications industry. So his Third Law might then state:

'For every regulatory action, there is an equal and opposite criticism'.

This often leads regulators to think that, if the howls of protest on each side of the debate are about equal, we've probably got the decision about right. So listening to informed criticism is an important part of our work and this book should assist in making such commentary more informed.

Peter Walker
Director of Technology
Oftel

March 2003

Preface

Regulation is a pervasive feature of the telecommunications services industry today. Government-appointed regulators and judicial or quasi-judicial bodies oversee it in countries at all stages of economic development. Its primary purpose is to encourage, nourish and maintain competition in national and international telecommunications services markets. It is, therefore, a fundamental feature of the legal and commercial landscape within which network owners and service providers operate. The modern development of regulation may be traced to the liberalisation of the industry from national monopolies from the 1980s onwards. There is an apparent paradox in that all this regulation arises to implement a process often described as 'deregulation'. 'Deregulation' refers to market liberalisation. 'Regulation' refers to the measures necessary to enable it, that is to permit new players to enter the market in the face of the power of the former monopolists to exclude them. The background is the immense technological progress of the late 20th century, combined with an underlying belief that competition results in better and more efficient services than does monopoly, central planning or government control.

This book aims to explain what regulation is, what regulators do and how they approach their task, why they do it, how they receive their powers, and how it bears on network operators and telecommunications service providers. It goes on through a number of case studies to show how regulators engage with topical issues, for example price control, interconnection, numbering, number portability and loop unbundling. This is a multi-disciplinary subject, drawing together economics (perhaps the primary motivator) and law and administration (the means of implementation) with business (the market players). Engineering and technology too have a role, since they determine the possible, contain the seeds of future development, and provide the means of solving problems that arise. This book is aimed at an intelligent reader from any of these spheres: engineers, technologists, lawyers, economists, business strategists and commercial managers. It is intended to equip readers to comprehend the work and pronouncements of regulators, to understand the language used, to engage critically in debate about regulatory matters and recognise how regulation affects their work.

I am conscious that this book probably breaks new ground. It presents a broad picture of regulation at a good level of detail, taking the middle territory between superficial overviews and the extensive, detailed literature emanating from the tributary disciplines of economics and law. It has proved an extremely interesting and fascinating task, one that has allowed me to use insights garnered over 30 years in the industry, firstly as a manager with a privatised BT and its predecessor the UK Post Office, and latterly as a consultant.

It is important to be clear not only about what this book is, but what it is not. It does not try to be, and cannot be, a detailed statement of the regulations that apply in any country at any given time. Legal frameworks and detailed rulings vary between countries, and change quite rapidly over time within each country. One of the standard textbooks for lawyers on this subject is published in a loose-leaf binding, purchasers being invited to subscribe for regular updates. The stress of this book, then, is on principles and not specific detail. If someone wants to know, say the current UK position on mobile price control, German licence conditions or the French rules on loop unbundling, he or she must look to the relevant regulator's published materials. Nonetheless, a reader of this book should be able to know what to look for, to discern the key issues and to understand what he or she finds.

This book is aimed at an international audience. The emphasis on principles as opposed to specifics means that it is not confined to regulation in one particular country. I have outlined and compared the regulatory landscapes of a number of different countries. It is true that I have drawn many (though not all) of my case studies from the national market I know best, the UK. This is a sensible choice, since UK practice has been moving consistently over recent years to the competition law principles of regulation that find expression in the latest (2003) European Framework described in more detail in the Appendix.

A number of the chapters engage with engineering and technology issues. I thus faced a decision in each case as to how much technology I should include, and in what depth I should describe it. I have used two basic principles in making these choices. The first is, 'To what extent is it necessary to know this technology to understand the regulation?' Interesting and magnificent though ADSL technology is, one can grapple with loop unbundling regulation while not understanding discrete multi-tone modulation. The second is, 'How readily can the technical knowledge be obtained elsewhere?' I have extended technical treatments where it seemed that even engineering specialists might not have the information at their fingertips. There is, thus, some detail about numbering and number portability, on which subjects little useful literature is readily available. Various descriptions of network architecture appear because the general reader might encounter difficulty in extracting them from technical works that usually go into much more detail. I decided to include a section on the ISO 7-layer model in the interconnection chapter (Section 5.8), after a solicitor had told me she needed awareness of it to do her work. It is far from intuitive outside the engineering world that one may need to regulate the part but not the whole of an interface between two systems. I have given a personal, high-level view of the technical and business rationale of next generation networks, in the hope of cutting through the fog of detail and of hype that surrounds this topic.

I owe an immense debt of gratitude to a number of people. Certain people have given me their time in interview, helping me get to grips with key issues and sort out the more from the less important matters. These include Alan Bell, Director of Strategy and Forecasting at Oftel, and David Newman, former Deputy Technical Director of

Oftel. As well as talking to me, Peter Walker, Director of Technology, Oftel, has read the manuscripts and helped me understand nuances that no amount of desk research would have gleaned. Charles Hughes, one of the series editors at IEE Books, has given me innumerable pieces of advice during the construction of this book. Roland Harwood, Commissioning Editor at IEE Books and Robin Mellors-Bourne, Director of Publishing, have both given encouragement during this project, and been willing and ready to help with various decisions about the direction this book should take. I thank them all, but hasten to add that what I have written is mine, and I take final responsibility.

On the central question, 'Has regulation been a success?', I sit on the fence in this book, though I touch upon it mainly in Chapter 10. A bigger question is whether the triangle of liberalisation, privatisation and regulation has improved the lot of humankind. That we now have more, better and cheaper services than we did 20 years ago is a matter of fact, but so are the downturn, the accounting scandals and the persistence of the access monopoly. History needs longer to run before we can convincingly tackle these questions. The revolution set off waves that have not yet come to rest.

John Buckley
Spring 2003

Acronyms

3G	Third Generation (when speaking of Third Generation Mobile Systems)
ACA	Australian Communications Authority (Australian regulatory body)
ACCC	Australian Competition and Consumer Commission (Australian regulator)
ACE	Automatic Calling Equipment
ADSL	Asymmetric Digital Subscriber Loop
ADSL Lite	A variant of ADSL
AGCOM	L'Autorità per le Garanzie nelle Communicazioni (Italian regulator)
ANATEL	Agência Nacional de Telecomunicações (Brazilian regulator)
ANFP	Access Network Frequency Plan
ANSI	American National Standards Institute
API	Application Programming Interface
ART	Autorité de Régulation des Télécommunications (French regulator)
AT&T	American Telephone and Telegraph Corporation (US company)
ATM	Asynchronous Transfer Multiplex
Austel	Australian Telecommunications Authority (former Australian regulator)
BS	Base Station (within GSM architecture)
BSC	Base Station Controller (within GSM architecture)
BT	BT plc is the UK former incumbent monopolist
BTA	Basic Telecommunications Agreement (of the World Trade Organisation)
CAMEL	Customised Applications for Mobile Network Enhanced Logic
CCBS	Call Completion to Busy Subscriber
CDSL	Consumer DSL
CEN	Comité Européen de Normalisation (or European Committee for Standardisation)

CENELEC	Comité Européen de Normalisation Electrotechnique (or European Committee for Electrotechnical Standardisation)
CLEC	Competing Local Exchange Carrier (a US local service operating company competing with the former monopolist)
CLI	Calling Line Identity
CMT	Comisión del Mercado de las Telecomunicaciones (Spanish regulator)
CPI	Consumer Prices Index
CPS	Carrier Pre-Selection
CRTC	Canadian Radio and TV Commission (Canadian regulator)
CSI	Customer Site Interconnect
DLE	Digital Local Exchange
DMSU	Digital Main Switching Unit (UK term for trunk or tandem exchange)
DNIC	Data Network Identification Code
DNS	Domain Name Server
DSL	Digital Subscriber Loop
DSLAM	DSL Access Multiplexer
DTI	Department of Trade and Industry (UK government department)
DTMF	Dual-Tone Multi-Frequency (method of operation of push-button tone telephones)
DWDM	Dense Wavelength Division Multiplex
E1	A basic 2 Mbit/s link
EBC	Element Based Charging
EC	European Commission
EEC	European Economic Community (old title of European Union)
ENUM	A protocol for representing telecommunications numbers within Internet Protocol addresses
E-OTD	Enhanced Observed Time Difference (method for locating mobile handsets)
ETR	ETSI Technical Report
ETSI	European Telecommunications Standards Institute
EU	European Union
EZ-DSL	A proprietary variant of ADSL
FAC	Fully Allocated Costs
FCC	Federal Communications Commission (US regulator)
FITCE	Federation of Telecommunications Engineers of the European Community
FRIACO	Flat Rate Internet Access Call Origination
FTTH	Fibre To The Home
FTTK	Fibre To The Kerb
GDP	Gross Domestic Product
GMSC	Gateway Mobile Switching Centre (within GSM architecture)
GNP	Geographic Number Portability (i.e. for 'ordinary' telephone numbers)

GSM	Global System for Mobile
HDF	Handover Distribution Frame
HDSL	High bit-rate DSL
HDSL2	Enhanced version of HDSL
HHI	Herfindahl-Hirschmann Index
HLR	Home Location Register (within GSM architecture)
HMSO	Her Majesty's Stationery Office (UK publisher of official documents)
IAD	Integrated Access Device
IBI	In-Building Interconnect
ICSTIS	Independent Committee for the Supervision of Standards of Telephone Information Services (UK)
IDA	Info-Communications Development Authority (Singaporean regulator)
IDSL	ISDN DSL
IEC	Interconnect Extension Circuit
IEE	Institution of Electrical Engineers (UK)
IEEE	The Institute of Electrical and Electronics Engineers (USA)
IETF	Internet Engineering Task Force
ILEC	Incumbent Local Exchange Carrier (a US local service operator being a former monopolist)
IN	Intelligent Network
INA	Individual Number Allocation
INCA	Interconnect Call Accounting
IP	Internet Protocol
IPR	Intellectual Property Rights
ISDN	Integrated Services Digital Network
ISI	In-Span Interconnect
ISO	International Standards Organisation
ISP	Information Service Provider
ISR	International Simple Resale
ISUP	ISDN User Part (of the Common Channel Signalling System No 7)
ITU	International Telecommunications Union
ITU-T	International Telecommunications Union, Telecommunications sector
IXC	Inter-Exchange Carrier (a US long distance carrier)
LE	Local Exchange
LEC	Local Exchange Carrier (a US local service operating company)
LLU	Local Loop Unbundling
LNP	Local Number Portability
LRIC	Long Run Incremental Cost
LRN	Local Routeing Number
MAP	Mobile Application Part (of the Common Channel Signalling System No 7)

MC	Marginal Cost
MCI	Microwave Communications Incorporated (US company)
MDF	Main Distribution Frame
MDSL	Moderate rate DSL
MEA	Modern Equivalent Asset
MFJ	Modified Final Judgement (a key decision in US deregulation)
MGCP	Media Gateway Control Protocol
MNC	Mobile Network Code
MNO	Mobile Network Operator
MNP	Mobile Number Portability
MR	Marginal Revenue
MSC	Mobile Switching Centre (within GSM architecture)
MSISDN	Mobile Station Integrated Services Digital Number
MSRN	Mobile Subscriber Roaming Number
MVL	Multiple Virtual Line (a DSL variant)
MVNO	Mobile Virtual Network Operator
NANC	North American Numbering Council
NANP	North American Numbering Plan
NANPA	North American Numbering Plan Administrator
NCOP	Network Code of Practice (UK regulatory guideline)
NGN	Next Generation Network
NGNP	Non-Geographic Number Portability
NICC	Network Interoperability Consultative Committee (UK joint industry – regulatory body)
NIPP	Network Information Publication Principles (for BT network information)
NPDS	Network Performance Design Standards (UK regulatory guidelines)
NRA	National Regulatory Authority
NTE	Network Terminating Equipment
ODTR	Office of the Director of Telecommunications Regulation (Irish regulator)
OFCOM	Office of Communications – to replace Oftel as the UK regulator
OFTA	Office of the Telecommunications Authority (Hong Kong regulator)
Oftel	Office of Telecommunications (UK Regulator)
ONP	Open Network Provision
OPTA	Onafhankelijke Post en Telecommunicatie Autoriteit (Dutch regulator)
OSI	Open Systems Interconnection
PARLAY	An open multi-vendor consortium to develop technology independent APIs
PBX	Private Branch Exchange
PDH	Plesiochronous Digital Hierarchy
PFI	Private Finance Initiative

PIN	Personal Identification Number
POC	Point of Connection
PON	Passive Optical Network
POTS	Plain Ordinary (or Old) Telephone Service
PPC	Partial Private Circuit
PSAP	Public Service Answer Point
PSBR	Public Sector Borrowing Requirement
PSD	Power Spectral Density
PSTN	Public Switched Telephone Network
PTS	National Posts & Telecommunications Agency (Swedish regulator)
PUC	Public Utility Commission (US state regulator)
RADSL	Rate-Adaptive DSL
RBOC	Regional Bell Operating Company (US local exchange operator)
RegTP	Regulierungsbehörde für Telekommunikation und Post (German regulator)
RPI	Retail Prices Index
RQV	Running Quality Value
SAC	Stand-Alone Costs
SATRA	South African Telecommunications Regulatory Authority (South African regulator)
SCP	Service Control Point (within the Intelligent Network Architecture)
SDH	Synchronous Digital Hierarchy
SDSL	Symmetric DSL
SHDSL	Single line HDSL
SIM	Subscriber Identity Module (identifying card in mobile handset)
SMP	Significant Market Power
SMS	Service Management System (for US number portability)
SMS	Short Messaging Service (of GSM)
SPC	Stored Programme Control (of telephony switches)
SRF	Signalling Relay Function
SSP	Service Switching Point (within the Intelligent Network Architecture)
SS7	Signalling System No 7, a widely used inter-exchange common channel signalling system governed by ITU-T standards
ST-FRIACO	Single Tandem FRIACO
T1	A basic 1.5 Mbit/s link
TC	Total Cost
TMT	Telecommunications, Media and Technology
TRAI	Telecommunications Regulatory Affairs India (Indian regulator)
TV	Television
UADSL	Universal ADSL
UK	United Kingdom
UPT	Universal Personal Telephony

URL	Universal Resource Locator
US	United States
USA	United States of America
USB	Universal Serial Bus
USO	Universal Service Obligation
VC	Variable Cost
VDSL	Very High Bit Rate DSL
VLR	Visitor Location Register (within GSM architecture)
VLSI	Very Large Scale Integration
VNO	Virtual Network Operator
VoIP	Voice over Internet Protocol
VPN	Virtual Private Network
WTO	World Trade Organisation
X25	ITU-T standard for packet communications

Chapter 1

Competition and privatisation: the background to regulation

1.1 Overview

Regulation of the telecommunications services industry is taking place in many countries of all stages of economic development. This follows from governments' policies from the 1980s to create competitive markets in this industry, where before there had been monopoly supply. Often, the national government owned the monopolist, which was in some cases part of the civil service, that is, of the government itself. Regulation forms one aspect of a restructuring process that has the three components shown in Figure 1.1: **competition, privatisation** and **regulation**. Any one of these could be implemented without the others, though obviously they must inter-relate when applied together. The motivation for this restructuring is a belief by governments that competition and private ownership will increase and continue to increase the power of the industry to satisfy growing customer demands with greater economic efficiency. The long-run effectiveness of all this remains to be confirmed by experience, through both boom times and lean times. Although rapid technological development has played a major role in reducing prices in this industry, there can be no doubt that additional strong and downward price trends have been observed over the last two decades in liberalising countries. There is, therefore, positive evidence for

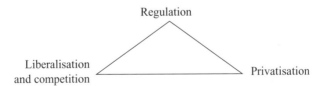

Figure 1.1   *The components of late 20th century telecommunications market restructuring.*

believing that competitive restructuring will prove a lasting success. Competition and privatisation are therefore here to stay, and a degree of regulation will be a necessity for some time to come.

Market liberalisation is the economic aspect of the changes engulfing the telecommunications services industry. It is the opening of national industries in monopolistic supply to new entrants, in the hope that newcomers will come forward and become part of a competitive market. Most current economic theories hold that a competitive market, always provided that it is possible to obtain one, will be much more efficient than a monopoly market. Efficiency in an economic sense implies that the physical, human and financial resources available to an economy are deployed to the maximum possible benefits and well being. The highest efficiency, or optimality, is proven when it can be shown that there is no possible reallocation of resources that could make someone better off while at the same time leaving no one else worse off. While perfect competition, a theoretical abstraction, may not always be attainable, oligopolistic competition (with a few though not many suppliers) generally produces better economic outcomes than does monopoly.

Privatisation is the political part of the equation. It follows from the proposition that privately owned providers have greater dynamic efficiency than suppliers under government ownership. This enables them to serve their customers with better products at lower costs, so raising economic efficiency. This proposition is contentious, and both it and the opposite may be held as a tenet of political conviction. Both affirmative and negating evidence may be found in specific situations. The main 'in principle' reason why privatisation might be beneficial comes from the operation of the capital markets. Publicly quoted companies in developed economies have to publish open, transparent, audited accounts, which are for all to see before they commit investment decisions[1]. Privatisation may in any case be desirable, simply to create a level playing field for the new players who will compete with the former monopolist.

Regulation, the final corner of the triangle, is the implementation dimension that facilitates change. A monopoly marketplace is highly unlikely by itself to evolve to the competitive market structure that gives the best service to the user. This is because an established, incumbent monopolist with high sunk investment presents a formidable barrier to market entry. A regulator has the delicate task of intervening in the market sufficiently to create the conditions for competition to start and flourish, but ideally not more than this. Regulatory intervention is best described as a corrective for market failure, having as its ultimate goal a sufficiently competitive market that requires no intervention. This is because competitive markets, once established, in principle provide the best economic outcomes by their own functioning. The quantity and nature of regulatory activity, and indeed the very need for *sector-specific* regulation of telecommunications, are valid and lively issues for debate. Different countries have adopted different solutions to the function of regulation.

The following two sections of this chapter address two main issues. The fundamental reasons why competition is (or may be) better than monopoly are explored. The advantages and disadvantages of private ownership are then compared with public ownership, and the process of privatisation is finally outlined. The role and work of the regulator are described in some detail in the remaining chapters.

1.2 The economics of markets

1.2.1 Market economies

Market economies are founded on the principle that economic efficiency and personal well being will be at their highest when individuals are free to follow their own interests as buyers and sellers, and where transactions are enacted between free and willing parties. The economy thus functions as a complex system of independently interacting agents. Such a system can be self-optimising in that it produces a stable equilibrium of supply, consumption and prices that is optimal. Common alternatives to market economies are **traditional economies** and **command economies**. In a traditional economy, typical of tribal societies, economic affairs follow a customary pattern while local leaders have the ultimate say in matters of allocation. This pattern works well in static societies, but may fail to respond to innovation and a changing environment. In a command economy, a powerful agent (usually the government) determines what people will produce, earn and consume. Twentieth century experience shows that this model fails partly because of the intractable complexity of the control task and partly because of its vulnerability to managerial corruption.

Market economies are tools that work well to achieve desired outcomes. They are, however, morally neutral in that it is people, acting both as individuals and through elected governments, that choose the outcomes they want. If a society wants, for example, to sustain the poor and care for the sick, people will present the demand and markets can deliver it. The 'hidden hand', a notion introduced by the 18th century social philosopher Adam Smith to portray the self-optimising property of markets, has no supernatural or theological reality. The 'hand' refers not to a *normative* principle dictating what should happen, but to a *positive* principle describing what does happen. Markets can succeed, often spectacularly so, but they may also fail and at this point governments intervene to correct market failure. This is what regulators and other institutions do, and by blending command measures with marketplaces they create the **mixed economies** common in advanced nations today.

The following sections examine basic theories of supply and demand in perfectly competitive and non-competitive markets. These serve not only to explore the reasons for liberalisation, but also to introduce economic concepts that will be used elsewhere in this book. This treatment is necessarily brief, and those who require a more thorough exposition are referred to a standard economics textbook, for example References 1 and 2. Perfect competition is a theoretical concept that provides us with useful insights, but is not necessarily an exact description of any part of the telecommunications services market. It is left to later chapters to explore specific situations that arise in the industry. In extreme summary:

- competitive markets determine an optimal equilibrium price for any product or service;
- price levels act as a signal for suppliers to change the level of production, and to enter or exit from the market;
- the most efficient suppliers are rewarded with the greatest profits, which (provided they do not make excessive dividend distributions) they can then invest in innovative products which grow the overall economy;

- capital markets allocate investments to the suppliers who will (or are expected to) make best use of them in creating value, that is growth and profit;
- monopoly supply is theoretically non-optimal in that a monopolist will produce at a higher price and smaller supply than the optimum. Nonetheless, monopolists can choose to exercise their market power in a benign way.

1.2.2   Demand economics

Consumers demand goods and services to satisfy their wants and needs. Normally, the less the price of a good, the more of it will be bought. The demand curve relates the quantity purchased (q) over a fixed period of time to the prevailing price (p), and economists call this entire relationship the **demand**. Changes of quantity purchased in response to price movements – in other words, moves along the demand curve – are not, therefore, changes of demand. An expansion or contraction of demand implies that the market has moved to a different curve, perhaps as a result of changes of income, income distribution, demographics, lifestyles or public taste. A typical demand curve is shown in Figure 1.2.

An important parameter is the **price elasticity of demand**, η. This measures the responsiveness of quantity purchased to price, and is normally negative. Defined in Equation 1.1, elasticity is clearly related to the gradient of the demand curve but is not the same, as it relates the proportionate quantity change to proportionate price change. Elasticity does not have to be the same over the entire demand curve and usually varies over it. Unit elasticity (that is, -1) implies that incremental changes of demand with price are neutral with respect to expenditure (or revenue), because an increase or decrease in price results in an offsetting change in the amount purchased. **Inelastic demand** occurs when the absolute value of η is less than one, in which case

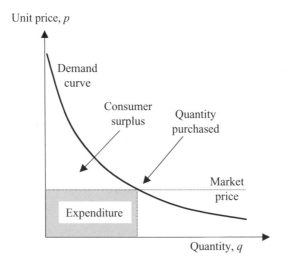

Figure 1.2   Typical demand curve.

a price rise will yield more revenue because the fall in quantity purchased is less than balancing. **Elastic demand**, with an absolute value of η above one, shows vice versa increased incremental expenditure when price falls.

Equation 1.1 Definition of price elasticity of demand

$$\eta = \frac{dq}{dp} \cdot \frac{p}{q} \tag{1.1}$$

Demand may be simply explained by the **utility theory of demand**, which provides useful insights into the behaviour of consumers. An individual consumer places a value on the benefit or satisfaction obtained from a purchase. This, the **utility**, can be expressed in monetary units, and our consumer will make a purchase provided the price is less than or equal to this. Few of us, of course, articulate this behaviour explicitly. The **principle of diminishing utility** holds that the benefit of successive purchases of extra units, or the **marginal utility** of each unit purchased, decreases as more units are acquired. The quantity purchased is then determined by the point at which marginal utility is the same as the price paid. The consumer does not purchase beyond this level, because an added purchase would give less benefit than the money sacrificed, or in other words would make him or her worse off. It follows that the majority of purchased units are bought at a lower price than the consumer would have been willing to pay. The cumulative difference, the area under the demand curve less the expenditure, is known as the **consumer surplus**, representing the added value to humankind of these economic transactions.

Indifference theory provides deeper insights into consumer behaviour than simple utility theory. Whereas utility theory treats purchases of each item (bread, petrol, telephone calls, etc.) in isolation, indifference theory recognises that most consumers have a finite budget, and so must make choices between the various things their money could buy. An **indifference curve** is a line on a graph relating the purchased quantities x and y of two products, joining all purchase combinations that yield equal satisfaction. In moving along an indifference curve, a consumer substitutes one product for another without gain or loss of benefit, so deriving constant utility. There are an infinite number of curves, one passing through every point on the graph and each curve having a different total utility. It is possible to construct theoretical indifference curves showing substitution between one particular product and the rest of the general shopping basket.

A consumer's self-maximising behaviour will be to select a basket of goods that places him or her on the indifference curve of highest total utility consistent with the budget. This selects one curve and a point on it, that is, it determines the quantities purchased. When the price of one item changes, our consumer moves to a point on a different indifference curve and thus a different balance of quantities. The change in quantity purchased may be analysed into two components. The first is **income effect**: the consumer becomes effectively better off (for a price reduction) or worse off (for a price rise) in real terms, so enabling a move onto a higher total satisfaction indifference curve (for a price fall) or forcing descent to a lower one (for a price rise). The second is

substitution effect, reflecting the revised purchasing balance that would have applied on the old budget but with the new relative prices.

When applied to real-life examples, indifference theory justifies the demand curves seen in practice, and explains anomalous behaviours. A well-known anomaly is the phenomenon of **Giffen Goods**, which display a regressive demand curve, where consumption falls with lower prices. These are inferior goods. When they become cheaper, income effect is negative and predominant, so the consumer finds that their cheapness frees money to purchase the better goods that he or she would have preferred before but for their lack of affordability.

Consumers maximise the total value of their basket of purchases by ensuring that equal utility is derived from the last unit of money spent on each. This makes their selection optimal, since no substitution of expenditure from one item to another can raise total utility.

1.2.3 Supply economics

Suppliers provide goods and services for consumers in the hope of making a living and a profit. Their price characteristic is opposite to that of consumers: they will supply more of a given good the higher its price. They extract revenue from sales of the product but must incur costs in producing it. A supplier's costs reflect the resources that must be consumed to produce output, and these are of four types.

- Physical resources, sometimes technically known as **land**, include land, water, energy, raw materials and other such things. These are nature's free gifts to humankind.
- **Capital** refers to the financial resources invested in setting up the enterprise and in providing working capital.
- **Labour** is the input of human effort, including research and development and other intellectual input.
- **Intermediate inputs** derive ultimately from the first three, and are items purchased in a partly finished form from suppliers of these goods and services.

The first three inputs are known as the basic **factors of production**. Suppliers must purchase them, and do so in markets where they are the purchasers. Suppliers can substitute between factors, for example by investing in more machinery or complex management systems to save human labour. A supplier seeks to optimise its factor consumption by ensuring that the last units of money spent on each have equal marginal productivities, so that there is no substitution that would make it better off by allowing it to produce at lower cost. The indifference curves that join equally productive combinations of factor inputs are known as **isoquants**. A supplier would, for example, substitute capital for labour in the face of a rising wage bill.

Costs may also and importantly be classified into **fixed costs** and **variable costs**. A fixed cost is something the supplier cannot change, a cost incurred simply by existing in its present form, even should it choose not to trade. Variable costs are incurred per item of output produced, and are correspondingly avoided when an item is not produced. The demarcation between fixed and variable costs depends on the timeframe the observer has in mind. From day to day, the only variable costs may be certain raw materials and the overtime element of labour. On a month-to-month basis, a supplier can vary its labour costs to a greater degree by recruiting or shedding people, or by refinancing a loan. On a year-to-year basis, it can replace machinery, build a bigger factory or move to a smaller one. In the very long run, all costs are variable.

Short-run variable costs for any one supplier arise in the context of a fixed amount of capital, for example a given factory or amount of plant. This plant usually has a baseline production capacity determined by the capabilities of current technology and prevailing market prices. As the supplier increases its output from zero to take up this capacity, it operates on a **falling cost curve** where the **marginal cost of production** is less for each additional item produced. The **average variable cost** per item produced, and indeed the **average total cost** per item (which adds in fixed costs), falls with increasing production as the supplier reaps **economies of scale**. Economies of scale, however, eventually come to an end, and **diseconomies of scale** begin above a certain level of production. At a simple plant level, this arises because machines and people are being stretched or production subcontracted; more generally for an industry there may be increasing management and communication overheads in maintaining bigger operations. As diseconomies of scale set in, the marginal cost of production passes a minimum and begins to rise. The average variable cost per item reaches a minimum at the production level where variable costs equal marginal costs, and begins thereafter to rise with increasing production. Figure 1.3 below illustrates these cost characteristics graphically.

In a perfectly competitive market, the supplier will produce up to the quantity where marginal cost equals market price; beyond that it loses money on each additional item produced, while below it there is still more money to be made. The qualification of perfect competition is necessary to set boundary conditions, one of which is that in a perfectly competitive market, all suppliers (and consumers) are small in relation

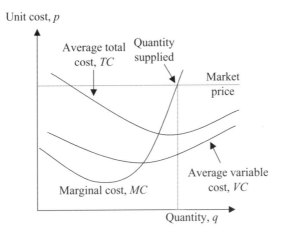

Figure 1.3   Typical supplier cost curves.

to the total market, so none of them can materially affect the total supply or demand in that market. Under these assumptions, our supplier is completely free to choose how much to produce, but has to accept the prevailing market price as a given. Note that the units sold, excepting the last, are sold above the marginal cost of producing them, and the cumulative excess is known as the **producer surplus**. The existence of a producer surplus does not necessarily imply profitability, as this depends on the overall level of costs, including fixed costs. A supplier would not produce were the market price below variable costs, as each extra unit produced would make losses worse. It would, however, produce above variable but below total costs, since each unit sold under those conditions produces at least a contribution to the unavoidable fixed costs.

Supplier profit per item produced is the amount by which market price exceeds the average total cost. These profits provide the return to the original investor or investors, and should be analysed into a **normal profit** element and **economic profit** (sometimes known as pure profit). Normal profit represents the opportunity cost of the capital tied up in the company, and is set by stock markets in accordance with prevailing rates of interest plus a risk premium characteristic of that industry. Investors will not invest in an enterprise making less than normal profits, because they could get a better return by putting their money elsewhere. Economic profit is the remaining excess after discounting the normal profit element. The existence of economic profits encourages newcomers to enter a market and supply it, with the result that under perfect competition, economic profits are eventually competed away.

The supply curve for one supplier, relating the quantity of product supplied to the market price, is basically its marginal cost curve. For a whole industry, it is the summed cost curves of all the suppliers in the market. Each supplier's cost curve will represent its cost characteristics at its chosen operating size, and many suppliers' cost curves plotted together will show as an envelope the long run supply curve of the whole industry as shown in Figure 1.4. This would change over time as a result of technology innovation, or a shift in the price of a key input factor.

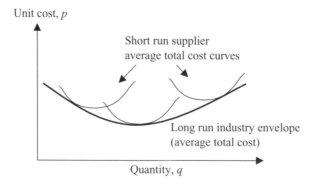

Figure 1.4   Short and long run cost curves.

1.2.4   The price system under perfect competition

Markets in perfect competition have an equilibrium price p_0 and quantity of supply q_0 for a given good determined by the intersection of the market demand and supply curves, as shown in Figure 1.5. At this point, there is neither unsatisfied demand, nor unsold production. The figure shows p_0, q_0, the money changing hands $p_0 q_0$, and the consumer and producer surpluses. A perfectly competitive market under price equilibrium is optimal in the sense that it maximises the sum of consumer and producer surpluses, as can be seen by inspecting the curves under displaced p and q values around the equilibrium. Note that displacements in certain directions imply behaviours that consumers or producers would not do voluntarily.

A market with **allocative efficiency** has no potential reallocation of resources that could produce a different bundle of goods that would make someone better off while at the same time leaving no one worse off. Combined forces operate in perfect markets to promote allocative efficiency.

- The price mechanism maximises the sum of consumers and producer surpluses.
- Consumers optimise their basket of purchases to maximise their utility.
- Producers optimise their profits by productive efficiency, that is by minimising costs or in other words resource consumption.

When a given market is in disequilibrium, there is either excess supply or excess demand. In either case, competition in the marketplace operates to restore equilibrium. In the case of excess supply, suppliers will compete for sales, thus lowering prices. As a result, suppliers will produce less and some may decide to leave the market. This adjustment may take some time, since suppliers often have assets that cannot be redeployed for other purposes, so they may shrink production only by failing to replace assets when they fall due. Under excess demand, consumer pressure bids up prices

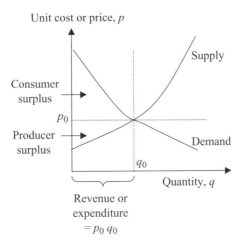

Figure 1.5   Market price under perfect competition.

to levels that some consumers are more than willing to pay. This in turn encourages both existing suppliers to produce more and new suppliers to enter the market, in the long run competing away the economic profits that build up during excess demand. In practice, many markets in enterprise societies are in ongoing disequilibrium as they react to changing circumstances brought by technological innovation, consumer wealth and other events. The opportunity to reap economic profit and the hope of gain drives the processes of innovation and adaptation.

1.2.5   Market structures and market failure

The perfectly competitive market, which generates productive and allocative efficiency as described above, is a theoretical abstraction and is dependent on certain assumptions.

- *Product homogeneity*   All suppliers sell an identical product. Such products are sometimes known as **commodity products**.
- *Perfect knowledge*   Customers have perfect knowledge of the products being supplied and their prices.
- *Lack of individual market power*   All suppliers and consumers are sufficiently small relative to the total market that none can materially affect the supply or demand in that market. Under this assumption, consumers can buy as much or as little as they like, while sellers have similar discretion over quantity of output, but neither has power to affect the ruling price.
- *Freedom of entry and exit*   Suppliers can choose at any time to enter or leave the market.
- *Existence of scale economies within the structure*   Each supplier's minimum cost output level (its capacity) is small in relation to the whole market.

Many everyday markets do not possess these features, giving suppliers varying amounts of **market power** to set price levels. The characteristics of perfect competition may be observed in certain markets, however, such as world commodity markets and capital markets (money markets and stock exchanges). Other market structures found in practice include the following.

- *Monopoly*, where there is only one supplier within a market.
- *Monopsony*, the counterpart of monopoly, where there is only one consumer within an industry.
- *Monopolistic competition*, where a number of competing suppliers provide similar but differentiated products, for example branded designer clothes. The market has some characteristics of monopoly in that loyal or committed customers can obtain a given product from only one source, though this is offset by the relative ease of consumer substitution and of market entry.
- *Oligopoly*, or oligopolistic competition, where a market has competitive supply but with a limited number of suppliers. Each has sufficient market share to give it some, though not unlimited, power to set prices. Oligopolists can act strategically by predicting and gaming one another's moves. They may collude to act like a combined monopoly, though many countries' competition laws make this illegal.

Market failure may be said to exist whenever a market fails to optimise productive efficiency and provide allocative efficiency, as would result under perfect competition. Optimality does not then exist, and there will be one or more redeployments of resource that could make at least one person better off without simultaneously making anyone worse off. The causes of market failure are of high relevance to any regulatory activity, and these will be explored in greater detail in the next chapter. One of the most significant market failures within the telecommunications service industry is that of monopoly, to which we now turn.

1.2.6   Monopoly

A monopoly market structure exists when there is only one supplier to a given market. This structure has been historically the case in the telecommunications services markets of most countries. To survive in the long run, a monopolist has to be able to exclude market entry by other suppliers. It is these entry barriers, which would not exist under perfect competition, that prevent the market from competing away excess profits or remedying inefficient production. Entry barriers may be economic or political in origin.

A frequent market entry barrier and cause of monopolistic supply arises from unused **economies of scale**. If current technology provides ongoing scale economies up to levels of production where a single supplier could supply a whole or large percentage of a market, then at the current level of demand the market will allow no more than one supplier to produce at the minimum point on the cost curve. Under these conditions, there is said to be a **natural monopoly**. Sometimes it is so clearly undesirable to duplicate a supply infrastructure, for example of water supply pipes or the road network, that the existence of a natural monopoly is obvious without detailed analysis. Where a natural monopoly exists, then it is not possible to obtain a competitive market, or at least not one that will generate allocative efficiency. In that case, a government can choose to live with the monopoly and possibly regulate it, or try to enforce wasteful competition. Even in a market that is not a natural monopoly, an established **incumbent** monopolist presents a formidable entry barrier to potential rivals, by reason of the high costs necessary to build up a production scale, brand image and distribution network that bears comparison with that of the monopolist. Whether or not certain parts of the telecommunications services industry are natural monopolies is a matter of lively debate to which we shall return in Chapter 4.

Besides economies of scale, a large supplier may possess **economies of scope**, and these can act as a barrier to market entry and so tend to reinforce monopoly. A supplier with **horizontal integration** produces more than one type of product, and in so doing splits its overheads such as research and development, sales and marketing and other corporate functions, so lowering overall costs. Another scope economy arises with **vertical integration**, where the supplier combines elements of the total supply chain from raw materials to retailing and distribution. This internalises transactions that would otherwise have taken place between independent agents in open markets, hopefully lowering their costs. Integration can lead to diseconomies of scope, however, should the management task become too complex to perform efficiently, or when an internalised operation could have lower costs if placed with a more specialised supplier able to serve a wider customer base.

The law may enforce monopoly, for example with the statutory monopolies enjoyed historically by postal and telecommunications authorities. Governments typically create statutory monopolies in order to take full control of industries that are seen as so socially or strategically essential that they are not willing to risk the uncertain outcomes that a competitive market might produce. Other forms of legal protection include regulated entry, for example the licensing of authorised practitioners in banking, law and medicine. Patent protection allows the owner of the patent to enjoy monopoly profits for a certain time as a reward for innovation.

It can be shown by straightforward theory that a monopolistic market will not generate allocative efficiency. This arises because a monopolist performs a fundamentally different optimisation from that of a supplier in perfect competition. The latter must take the market price as given but can choose its level of supply. The monopolist can name the price, but having done so must face the demand curve over which it has no control and which will decide its quantity of supply. This leads it to a different equilibrium for profit maximisation. Whereas the competitive supplier finds its supply level by equating marginal cost with price, the monopolist equates marginal cost with marginal total revenue. The marginal revenue obtained by selling an extra unit will necessarily be lower than the price, since when the demand is (as normal) negatively sloping, it must lower the price of *all* units in order to sell the extra. These relationships are illustrated in Figure 1.6 below, here simplified with linear demand. It follows that the monopolist will supply less output at a higher price than would have applied under competition for a supplier with the same marginal cost characteristics. This behaviour has the effect of transferring some of the consumer

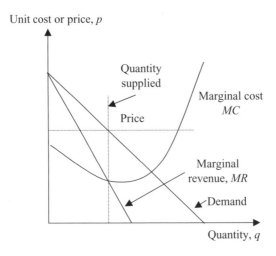

Figure 1.6   Monopoly profit-maximising equilibrium.

surplus to the producer surplus, but in any case lowering the sum of the two. This loss of the sum of the surpluses relative to what would have applied under competition is termed the **deadweight loss of monopoly**. None of this tells us whether the monopolist will be making profits nor how much, as this depends on its cost structure.

The final view we take of monopoly rests on weightier matters than a theoretical analysis of profit maximisation. On the one hand, the lack of competitive threat gives a monopolist's management and unions reduced incentive to provide a cost-efficient service. The fact that BT could in the 1990s expand its services, reduce prices and grow profits while reducing its overall size by more than half, demonstrates that cost inefficiency may be a significant feature of monopoly, although technological improvements undoubtedly played a part. On the other hand, monopolists can exercise their market power in benign ways, perhaps counterbalancing the economic evil they might at first seem. They may choose not to maximise profit for a variety of reasons. Monopolists can cross-subsidise, using excess profits obtained from one product to underwrite lower prices on others. This would be impossible in an open market, as competitors would simply undercut and capture the higher priced business. One example of cross-subsidy in telecommunications was (and is) the subsidy of loss-making line rentals and local calls by revenues from correspondingly over-priced international and long-distance calls. BT defended this practice before privatisation by arguing that it was valid to subsidise the line rental from call charges because each added line raised the value of the network to all its users. This is sound economics, because an added line does indeed benefit other users, even though it is not taken into account by the market price system. An economist would term this benefit a **positive network externality**, and argue that an appropriate level of cross-subsidy remedies the market failure posed by the externality[2]. The subsidy of local calls is a redistributive transfer between consumers, which might be justified in terms of welfare economics. Monopolists have the power to maintain a quality of service that might under competition tend towards the lowest standards practised in that marketplace. This power can be used both to the benefit of the consumer or to his or her detriment, should standards be maintained at a higher level than customers would naturally want, simply to sustain the monopolist's self-interested choices of activity, employment and turnover.

Monopolists making high profits have in some cases used them to enormous public benefit. For example, AT&T's Bell Laboratories invented the transistor, while BT built up an impressive pool of skilled people inside and outside its famed laboratories. Their contribution to the national economy has gone well beyond that reaped internally by BT and its predecessor, the British Post Office. Some voices argue that we should not take too negative a view of monopolies and oligopolies for two principal reasons. First, the hope of monopoly profits has spurred the innovations that have advanced the quality of all our lives immeasurably. Second, no monopolist can be entrenched in the long run because of the process of **creative destruction**, the bringing by technological progress of new products and processes that undermine the long-term persistence of entry barriers. Microsoft appealed to these basic principles in self-defence during anti-trust proceedings in 1999.

1.3   Privatisation of publicly-owned enterprises

1.3.1   The political basis of privatisation

Privatisation is the transfer of economic activities under public ownership to private ownership. Privatisation of telecommunications and other businesses became popular in the 1980s and after, in the conviction that private sector businesses perform more efficiently than public sector concerns. These supposed benefits are different from, and additive to, the gains resulting from the conversion of markets from monopolistic to competitive structures, and should not be confused with them. Privatisation has been seen in most countries as a pragmatic development, motivated by the desire for dynamism and efficiency. It has not been primarily motivated by political doctrine, although that has played a greater or lesser part in the expounding and justifying of privatisation programmes to electorates. A positive view of private enterprise has not been constant throughout history, and the UK's nationalisation (that is taking into public ownership) of key industries such as railways, coal and steel after the Second World War (1939–1945), was motivated also by hopes of greater efficiency. In those days, the private sector was perceived to have failed to generate necessary investment.

The political basis of privatisation is the notion that governments best maximise people's freedom and happiness by minimising their own activity. In other words, governments should only do what only governments can do, and 'big government' is, in principle, a bad thing. This figures in the radical libertarian thought of the 19th century, for example of British historian, philosopher and economist John Stuart Mill (1806–1873). In 'On Liberty' (1859) he wrote, 'The only purpose for which power can be rightfully exercised over any member of a civilized community, against his will, is to prevent harm to others'. Abraham Lincoln (1809–1865), 16th President of the United States, asserted, 'The legitimate object of government is to do for a community of people whatever they need to have done, but cannot do at all in their separate and individual capacities'. Thomas Jefferson (1743–1826), third US President and principal author of the Declaration of Independence, echoed this saying, 'That government is best which governs the least, because its people discipline themselves'. Jefferson hints here that there may be moral preconditions for liberty to succeed. These thoughts inspired the *laissez-faire* tradition in western politics, of leaving as much freedom as reasonably practical to individuals and markets. The opposite tradition, that of *dirigisme*, is much more content with control, prescription, centralisation and intervention as a means of achieving civic and economic goals.

1.3.2   Government spending and the public purse

The transfer of enterprises from the government to the private sector may be undertaken for positive reasons, to invigorate the national private sector, or with the negative aim of cutting the size of the public sector. The former is probably the chief driving force in developing former colonial countries, where, possibly lacking developed private sectors, they find privatisation an opportunity to seed an emergent business culture with enterprises readily able to generate profits given competent management.

Privatisation makes it easier to attract foreign investment, which may also bring strategic partnerships transferring technical and business expertise into the country. Investors may advance funds for private ventures when they would have found it very difficult to raise equivalent confidence in the national government. In developed economies, however, more usually the negative desire to diminish the public sector has predominated.

The size of government relative to the total economies of developed countries has grown inexorably throughout the 20th century, reaching in many cases 30 to 50 per cent and more of total economic activity (that is, of gross domestic product, GDP) from bases around 10 per cent. This is a cause for concern for several reasons. First, there is genuine doubt over the ability of a large public sector to promote allocative efficiency, or indeed to mirror people's real welfare. Second, large public sectors imply high rates of taxation, raising a fear of depressed incentives for individuals to innovate, take risks and work hard. Finally, rates of taxation in some countries are now so high, for example with marginal income tax over 50 per cent and value added tax (sales tax) greater than 20 per cent, that there may be little remaining scope to raise them further. Should a government exhaust its capacity to raise taxes, it would increasingly be constrained by unavoidable spending, so losing control of policy and weakening its ability to respond to disasters, war or depression.

Government spending in a developed country consists of the five basic elements listed below. Their cash and investment requirements compete for a share of government finance, both with one another and with any desire to cut or stabilise taxes. The total government demand for funds and hence the **public sector borrowing requirement** (PSBR) may run to billions or tens of billions of pounds. This is an amply large demand to affect interest rates in the money markets.

- Servicing of the national debt (interest).
- Transfer payments.
- Provision of 'free' services.
- Residual funding of traded services.
- Publicly owned industries (some of which may be profitable and so generative rather than consumptive of cash).

National indebtedness provides an essential buffer of borrowing, enabling a government to maintain a steady (or managed) flow of expenditure in the face of uncertain tax revenues, which vary with feast and famine in the economy. There is little a government can do in the short term to change the debt burden, though with responsible long term planning it can repay debt, and avoid irresponsible spending that the country cannot afford.

Transfer payments are redistributive sums taken from donors through taxation and given to recipients as social security payments, disability benefits, old age pensions, child benefits and the like. All developed economies have these, although the debate about amounts and eligibility is lively and will continue so to be. Transfer payments are defended in the name of social justice and moral compassion, and opposed for their possibly lessening the incentive for individuals to help themselves. Their economic impact is potentially positive first in that they address the costs of social exclusion imposed on the community by way of crime, public health risks and lost human opportunity, and second in bringing into economic activity people who might otherwise not have participated. Their effect is doubly negative when people having the option of contributing to the economy choose cushioned inactivity, both through the direct cost of the benefit and by the lost input.

Free services in a typical modern economy include roads, education, defence, policing and the administration of government. While free at the point of delivery, these services are very expensive and people are compelled to purchase them through taxation in the exact form that the government chooses. Normally, it would make no economic sense to insulate a consumer from the cost of something, as this would stimulate demand to a level at which the consumer's utility was well below the resource cost of the consumer good. Free services can be justified, however, when there is clear public advantage, the benefit is non-excludable and the price elasticity of demand is low. Non-excludability means that it is impossible to prevent from enjoyment anyone who might choose not to pay, so there can be no market for a non-excludable service. Demand is often inelastic for these types of service: a family with two children is unlikely to purchase four places at the local school on account of cheapness, nor do people seek unnecessary medical treatment when the price is right.

Residually funded traded services are those for which consumers pay at the point of use, but the government deliberately provides them at less than true cost and so incurs a loss on delivering them. Public transport is a well-known example, as are sports, cultural and leisure facilities. These can be justified when there is an element of public good, and they may incorporate a redistributive element. Subsidies can address externalities: a transport route adds to the value of the network even for those who do not traverse it; public transport may reduce road congestion and pollution while also providing insurance value for car-drivers against the day the car is unavailable. The user purchase price is not only to relieve the cost to government, but also to restrain excessive consumption, since most of these services have finite capacity. When a public enterprise makes a loss on a free or a residually funded service, this need not imply economic inefficiency. The joint purchasers are the end consumer and the government, who may together receive excellent or poor value for their money.

The privatisation of a publicly owned enterprise removes the pressure of its cash demand from government finances, making way for more spending on other services, for tax stability or tax reduction. While the removal of an industry from the public sector does not, of course, lessen its demand for cash on the economy as a whole (ignoring for the moment the relative efficiencies of public and private sectors), a private company's funding requirement has less impact on interest rates. This is because private companies have access to global equity capital markets, unlike public companies, which only use government bond markets. Indeed, private investors may be only too willing to invest in the industry. Profitable privatised companies pay corporate taxes, which may compare with or indeed exceed their profits (if any) when in public ownership.

1.3.3   Commercial aspects of privatisation

It has been an article of faith that the privatisation of government-owned concerns would make them more efficient companies. This section examines some of the commercial reasons why this might be so:

- privatised companies may be able to make decisions more quickly than in the government service;
- privatised companies have a greater range of freedoms to achieve their objectives.

On the other hand, some pitfalls have been argued to militate against the effectiveness of privatised industry:

- a public undertaking can obtain finance at lower rates of interest than most private companies;
- a private company is not free to pursue non-commercial, social objectives;
- a private company may inflict long term damage on a national economy should it pursue short-term profit maximisation at the expense of longer term stability and growth;
- private companies may take a more selfish view than public companies of patent and design rights, and use them without compunction to impede progress if they considered this to be in their interest.

It is a common observation that large public telecommunications companies make their decisions very slowly. Their typically hierarchical management structure is frequently blamed for this, although hierarchies can in fact move quickly when they so desire, the command and control structure having been first developed for armies. The real origins are probably cultural, as evidenced when these same companies remain lumbering giants a decade and more after privatisation! Nonetheless, any public sector organisation, whatever its culture, must obtain governmental sanction for major decisions, and this must inevitably and sometimes correctly make for delays while the political process operates.

The laws that first established them usually bind government-owned corporations, defining what they can and cannot do. To enter a new market or raise money in new ways might in these cases require primary government legislation. This removes freedoms a private company would have as a matter of course. Management given a wider range of options, while obviously and ultimately accountable to shareholders, will be in a stronger position to add value by enhancing growth and profitability. Typical freedoms of private companies that are not always available to the publicly owned company include the following.

- They can raise capital, including risk capital, in the equity markets, and transact home and foreign loans.
- They can enter into strategic partnerships and collaborations with other companies.
- They can participate in mergers and acquisitions.
- They can restructure themselves by consolidation, or by splitting into separate companies.

- They can give credit, or advance loans, to their customers.
- They can make speculative investments.
- They can experiment with new products and prices, and formulate special offers.
- They can develop new revenue streams, some of which may go beyond their original remit when public companies, as for example when an electricity utility sells gas, or a post office retails telephone services.
- They can operate outside their national markets, competing for a share of global markets.

It is true that most governments enjoy a lower cost of capital than private companies. However, this is an oversimplification and not normally a relevant observation. A private company's cost of capital is determined by prevailing interest rates, weighted by a 'risk premium' determined by the money markets' credit rating of that company, to offset the market's view of the failure probability of that firm or its industry sector. Governments have low or zero risk premiums because they very rarely go bankrupt. However, governments do bear risk premiums but in a different way. This is because risk sometimes eventuates and projects have to be bailed out with extra funding. Nonetheless, a poorly thought-out privatisation may force the community to bear the risk premium not once but three times: once through the private firm's cost of capital, twice if the terms of privatisation are set in favour of the private company to derisk its capital and profitability, and thrice should an essential venture face difficulties. A government cannot abrogate ultimate responsibility for an essential industry.

Private companies cannot be compelled to trade at a loss, so privatisation largely removes government freedom to manage an enterprise in pursuit of social policy objectives that conflict with commercial profitability. There are strong economic arguments that interventions distorting the behaviour of free markets make everybody worse off in the end, thus suggesting that privatisation is here reducing a freedom that a wise government should use only very sparingly. If a government decided that a subsidy was sound policy, then government ownership would not be essential as it could enter into contracts with a private provider of services. As later chapters will show, there is some limited opportunity for governments to impose social obligations on a private company by regulation. Examples in telecommunications include universal service obligations, geographically averaged prices that do not vary from region to region, unbalanced tariffs that preserve a line rental subsidy and interoperability between competing services.

Telecommunications services, in company with electricity supply, railways and other basic utilities, are absolutely fundamental to modern societies, so the concern that an irresponsible privatised company might maximise profit by weakening or destroying the essential fabric of the nation is a real one. Even an incompetent manager can show a short-term profit by depleting assets. There can be, of course, no guarantee against this happening under government control. The main ground for confidence that privatisation will not place key assets in the hands of foolish, selfish or criminal managers is the behaviour of the money markets and the investment community. Privatised utilities are very large companies, in which investors will be looking for sound growth opportunities with dependable earnings. In doing this they will invest in the managements that best demonstrate a track record in and understanding of the industry, allied with credible business plans that withstand scrutiny by analysts. This process is not without its dangers, for example when there is uncritical enthusiasm for fashionable strategies, but it does provide a supervision and control over companies. Some people believe that governments, subjected to political pressures, have made, and will make, worse decisions than the private sector would have done.

1.3.4 Cultural factors in privatisation

A culture of a company is its members' belief set that provides them with a model for understanding their purposes and roles within the company. It guides the behaviours they should adopt for survival, for happiness, and for advancement, and engenders an atmosphere determining their effectiveness and well being. Culture values can take a religious nature, and the term 'spiritual values' has been used in management circles when speaking of culture. The culture of a company is forged at the inter-personal interface of employees with their peers and with their superiors, though the effect of it is felt in all internal and external relationships. The finest and most pro-ductive cultures are ones where people act rightly, justly and respectfully with their colleagues, community and environment. In a company with a culture such as this, the brightest and best will be able to compete and achieve their highest potential, while all employees will feel valued and find satisfaction in hard work. Privatisation enthusiasts argue that private companies have superior cultures to public enterprises, making their people better and more effective people, thus helping private enterprises have greater dynamic efficiency.

The culture of a company is not easy to measure. Insiders possessing the most complete knowledge are the least likely to form an impartial view. Official statements of mission or company values are irrelevant, since it is how a company actually behaves that matters, not how it says it behaves. A company's culture can be assessed by observing: the lives of the men and women it most honours; the table-talk stories, legends and folklore that form the core of everyday conversation; and finally the way it treats its people of highest quality and capacity. Cultures may be described with phrases such as 'culture of excellence' or 'culture of machismo'. Table 1.1 lists a number of descriptors of culture, and a handful of these may apply in a given case. Large companies are not monochromatic, so one might find subcultures at different sites, in different functions, or at various management levels. Some cultural attributes are obviously good or bad, while others confer situational advantage or disadvantage. A few are neutral, giving a good company the freedom to have a characteristic personality.

Is there an intelligent warrant for believing that a privatised culture will be better than a public one? At one extreme, endemic corruption can compromise a public ser-vice to the point of ineffectiveness. On the other hand, a strong public service ethic can produce creditable and enviable results. The British National Health Service has been regarded as a model for others, at least until recent years of under-funding. A good private company, managed with fairness and goodwill, offers tremendous potential for employee motivation, where individuals can work hard and enjoy commensurate

Table 1.1   Descriptions of company cultures.

Academia	Dedication	Intimidation	Questioning
Accountants	Democracy	Laddishness	Quick fixes
Aggression	Determination	Lavish living	Racism
Amateurism	Dishonesty	Loyalty	Respect
Amorality	Dog-eats-dog	Luck	Safety
Analysis	Efficiency	Machismo	Sales and selling
Arrogance	Elitism	Meditation	Scapegoats
Autocracy	Empire building	Meetings	Sclerosis
Back-stabbing	Engineers	Nastiness	Self-sacrifice
Bad language	Excellence	Nepotism	Seniority
Blame	Exhibitionism	Novelty	Servant leaders
Bullying	Fairness	Oafishness	Service
Bureaucracy	Gaming	Over-	Sexism
Care for detail	Gambling	engineering	Short-termism
Chaos	Gentlemanliness	Paralysis	Spirituality
Cheating	Gerontocracy	Patronage	Straight talking
Committees	Get rich quick	Personality	Success
Competition	Goal-direction	Pettiness	Sycophancy
Conformity	Grandeur	Politics	Team sports
Consensus	Health	Power	The big project
Corruption	Hero-worship	Procedure	Theft
Criminality	Humility	Professionalism	Travel
Customer care	Hyper-activity	Profit	Workaholism
Debate	Image	Public service	
Deceit	Innovation	Pushfulness	

reward in achievement. Sadly, however, not all private companies are like this, and some have an egoistic management *Reich* that generates a tense climate, where hope and fear run uncomfortably hand-in-hand, averaging out in a dull mediocrity. Chaotic management styles *may* promote creativity, but more frequently they shelter (or sanctify) incompetence, ensuring that little is achieved beyond muddling through.

The long-term security of a publicly owned company makes possible the formation and execution of sound strategies and the building of a highly expert workforce. However, a stable revenue base and relative security, alloyed with feeble ethics and weak leadership, can remove the need for individuals to perform in any economic sense. This makes a fertile breeding-ground for pathologies such as cultures of laziness, procedure, over-engineering or the easy life. Management of such an enterprise may become a pure political power play, a culture that typically pulls forward managers who will find it comfortable to sideline people of ability and distinction. Wherever mediocrity is the norm, it is also the ideal. Highly talented individuals who successfully circumnavigate pitfalls such as envy, bullying, chaos, cynicism or unrealistic expectations, might still find it hard to evolve into truly world-class players.

This is because they may not get the stretching challenges needed for development, or because their leaders are not worthwhile role models. Ultimately, however, organisational culture depends primarily on the quality of the people in the enterprise, and especially of its leadership. It is difficult simplistically to identify superior and inferior cultures with private and public ownership.

1.3.5 The morality of privatisation

It has been argued that the privatisation of publicly owned assets is fundamentally immoral, most notably by former British Prime Minister Harold Macmillan[3] when he applied to it the metaphor of 'selling the family silver' [3]. The imagery was of the aristocratic heir who unwisely dissipates his family fortune. The message was that it is wrong, in principle, to dispose of human and physical assets built up over decades of endeavour at taxpayers' expense. However, this argument does not stand up when stated in such a simple form. When a government sells off a publicly owned asset and receives fair monetary value in return, the transaction is neutral. Whether an act of privatisation is moral or immoral depends on the valuation of the transaction, and the way the money is used after receipt.

A privatisation knowingly transacted below its true asset value would be an act of a foolish, bad or wicked government, who might do so for one of these reasons:

- to give advantage to well-connected people;
- to privatise at any cost, maybe for doctrinal reasons or to get capital spending off the public sector books;
- to privatise hastily, without proper thought and planning;
- to simulate the appearance of interest in a privatisation that had generated little real investor enthusiasm;
- to de-risk an enterprise or its profitability before sale to a hesitant buyer.

In the same category of bad government would be a value-destructive privatisation, for example by splitting up an industry to create artificial competition, or to render it difficult or impossible to renationalise at a later date. The mere fact that shares rise in value after privatisation is not in itself a sign of irresponsible valuation, as some utilities were very difficult, certainly in the early days, to value accurately. Share prices can also reflect technical and sentiment factors as well as fundamental value. The revenue received from a privatisation is a capital receipt and ought to be used as such, for example for infrastructure investment or paying off the national debt. To use it for consumption or to fund tax reductions would indeed be 'selling the family silver'.

1.3.6 Forms of privatisation

The simplest form of privatisation takes place when the government converts a publicly owned corporation in its current operational shape into a joint stock company. The shares, initially owned by the government, are then sold to investors who as shareholders become the new owners of the enterprise. This type of privatisation has been the most common in the telecommunications industry. The sale of shares may take two forms, a **public floatation** as typically used in Western Europe, and a **trade sale** such as seen in Central and Eastern Europe. Under a public floatation, the government and its advisers invite the investing community and general public at home and abroad to subscribe for shares at a stated offer price. If the issue is oversubscribed (as has commonly been the case), the government scales down individual applications when allocating shares. In the UK and elsewhere, the allocation has been weighted in favour of small investors. Under a trade sale, the government seeks strategic investors, or consortia of investors, for a bulk sale of shares. This is usually preceded by a tendering exercise where investors are invited to bid for the shares and submit their plans for developing the company. The government, who may have set development targets, awards the shares to the investor(s) who submit a competitive bid combined with an attractive and credible business plan. Typically successful consortia have included investing partners, perhaps banks or cash-rich utilities, together with foreign telecommunications operators who will be expected to invest and transfer expertise into the host country in return for a share in the profit stream.

In many cases, the privatisation of a monopoly telecommunications operator is completed only in part, when the government sells some of the shares but retains a large and perhaps a majority holding. There are various reasons why governments do this.

- National monopoly telecommunications operators are typically very large companies, often becoming their country's largest quoted company when launched on the stock exchanges. A floatation of such a company in its entirety would be too big a cash call for the investment markets to digest. In this case, the government usually completes the privatisation in two or more public offerings, or *tranches*.
- Dividing the floatation process into tranches may help the government obtain a better (or fairer) price for the company. Given that initial valuation is difficult, second and subsequent tranches will be informed by a market view of the value of the shares.
- Whatever its rhetoric, the government may not be wholly committed to privatisation and may wish to retain a large or majority stake under government control, while simultaneously securing the benefits of equity finance. The government may have done this for national political reasons, to please unions anxious to protect jobs, or from a desire to prevent control going outside the ruling party or the country.

Some telecommunication operators have a so-called 'golden share', enjoying special powers of veto over mergers, acquisitions, takeovers and foreign control, and which the government continues to hold even when privatisation is otherwise total. There is nothing iniquitous in the existence of such a share, provided its existence is made clear to all potential investors in the floatation prospectus. Some governments (such as the UK's) have given up their golden share to give their incumbent operator greater freedom to act as a global player. Foreign regulators might well be less willing or unwilling to consent to its acquisition of a slice of their national telecommunications market if they perceived the golden share to imply a favoured status in the home market.

Other forms of privatisation that have been less common in the telecommunications services industry include:

- enterprises restructured before floatation;
- outsourcing;
- private finance initiatives (PFIs).

Governments split some industries into separate component companies before privatisation, for example, electricity generation, gas supply, water supply and railways in the UK. The main motivation is to create some basis for competitive supply in industries where there are natural monopolies.

Full privatisation is not possible for all parts of the public sector because of the non-commercial nature of their business. Examples include defence, prisons and education. This need not, however, preclude the involvement of private providers, albeit under ultimate government control, as a means of reaping some of the benefits of privatisation. The private sector may be involved through **outsourcing**, or by a more complex **private finance initiative**. Outsourcing arrangements with private contractors may cover core activities such as medical care, or non-core functions such as catering, cleaning, telecommunications or computing services. Outsourcing may be used as a method of invigorating the efficiency of a service stalled in the hands of an inflexible public-sector workforce, although in many cases, in-house direct labour organisations have continued to provide services because theirs is the best proposition in fair and open competition. Private finance initiatives (see, for example, Reference 4) are more complex forms of outsourcing where a **special purpose vehicle**, a private limited company, is set up to finance, build and operate an operation such as a prison, hospital, school or railway in return for contracted remuneration during the lifetime of the service.

Privatisation is neither panacea nor instant solution to the problems of any industry, and each case needs to be carefully evaluated on its merits. It is most important that privatisation not be undertaken lightly or wantonly, simply because it happens to be a fashionable solution. A well thought-out privatisation can bring extra resources into play, widen customer choice, and increase value for money through specialisation and dynamic efficiency. A poorly thought-out privatisation may be little more than a very expensive way of getting capital spending off the government's books, and could in the extreme wreck an industry with grave costs to the whole community. The application of structural solutions to operational problems may rightly attract the criticism of 'fiddling while Rome burns'.

1.4   Regulation: a guide to this book

The remainder of this book describes the role and work of regulation in the telecommunications services industry.

Chapter 2 explains the rules and regulations typically applied by way of market intervention, and the reasons for those rules. Chapter 3 covers the legal and administrative frameworks that appoint regulatory bodies and give them the powers they need to function. It also explores the modes of operation of regulation, and some of its problems. Recognising that regulation is a constructive and potentially hazardous activity needing discretion in its execution, Chapter 4 addresses issues of regulatory strategy, including a review of the available methods of price control. The fine detail of regulation, of course, varies not only from country to country but also continuously with time. It is a living process. The purpose of this book is to explain the principles and practice of regulation. It does not, and cannot, provide a detailed statement of any one country's regulation at a point in time.

Chapters 5 to 9 analyse a number of specific issues upon which regulators take action. Many of these have an engineering dimension. These analyses have a threefold purpose: to explain the issues, to show how regulators have dealt with them, and to illustrate the conflicts, judgements and decision-making that are part of the regulatory activity. Technical material is covered in sufficient depth to allow professionals from engineering and other backgrounds to understand the issues. References are generally provided where required for those who wish to engage with technical aspects in greater depth. Readers requiring an overview of the general technical landscape of the telecommunications industry are referred to one of the general telecommunications textbooks, of which Reference 5 is an example.

Chapter 10 explores future trends in telecommunications regulation and how they may be affected by technology. The Appendix introduces the European Directives due to come into force in July 2003 to govern European regulation in the forthcoming years, and paraphrases their provisions.

1.5   Notes

1   Recent industry scandals show that learning processes continue in our societies.
2   There is a more detailed treatment of the concept of externalities in Chapter 2.
3   Harold Macmillan, 1894–1986 and later the 1st Earl of Stockton, was Conservative Prime Minister from 1957 to 1963.

1.6   References

1   LIPSEY, R. G., and CHRYSTAL, K. A.: 'Principles of Economics' (Oxford University Press, Oxford, 1999, 9th edn.)
2   BEGG, D., FISCHER, S., and DORNBUSCH, R.: 'Economics' (McGraw-Hill, London, 2000, 6th edn.)
3   Speech to the Tory Reform Group, 8th November 1985, reported in *The Times*, 9th November 1985
4   YULE, I.: 'The public face of a private finance initiative', *IEE Engineering Management Journal*, 2001, **11**, (3), pp. 115–21
5   FLOOD, J. E. (Ed.): 'Telecommunications Networks' (IEE Books, London, 1998, 2nd edn.)

Chapter 2

The task of regulation

2.1 Why regulate?

2.1.1 Mission and goals

The regulation of an industrial or business activity is a form of market intervention that aims to stimulate behaviours that would not by themselves emerge. It is a process of developing, agreeing, setting, evolving and enforcing rules of conduct and engagement. It is undertaken to encourage desirable outcomes, or to remedy proven problems. While there are many possible reasons for regulating an industry, the principal focus for contemporary regulation of the telecommunications services industry is the creation, nourishment and maintenance of competitive markets. Elsewhere, however, regulation sometimes relates to:

- preservation of market ethics (as in retail finance);
- maintenance of professional or technical standards (as in medicine, law, transport, manufacture and in the building and repair trades);
- consumer protection (as in health, hygiene, safety and trade descriptions regulation);
- safeguarding of a workforce (as in apprenticeship, training and exclusivity regulations).

Telecommunications regulation must serve government objectives for the telecommunications services industry. This industry is a fundamental and essential part of the infrastructure of a modern economy. A national telecommunications policy might typically have the following objectives. (As explained in Chapter 1, many countries now regard competition in the supply of telecommunications services to be a fundamental tool for meeting these goals.)

- A viable, up-to-date telecommunications industry that compares internationally with best practice.
- Universal availability of basic services, and wide availability to the business sector of advanced services.

- Cost-efficient and affordable services.
- A chosen degree of competition.
- Proper support of national security, law enforcement and defence requirements.
- A framework for setting prices, to protect customers from abuses of market power and provide incentives to improve efficiency.

2.1.2   Reasons for intervention

Regulatory intervention in a market is a hazardous activity that is capable of distorting markets as well as benefiting them. Distortion of the operation of free markets rarely improves total well being, and in many cases makes the generality of people very much worse off. Regulation must, therefore, be justified and appropriate, and shown to be so. It must address clearly understood problems. It is always right to ask the question, 'Why regulate?' and to review it periodically in the light of market changes including those brought about by that regulation. Where regulation to encourage competition succeeds in bringing that about, it should then cease, reduce or refocus its operation.

The primary reasons for regulatory intervention in the telecommunications services industry are twofold. First, there is **market failure**. These are phenomena that prevent the market from producing an optimal allocation of resources, and are described in Section 2.2. Second, in telecommunications certain universally available services may be provided at an affordable price on a non-commercial basis for reasons of national policy and social well being. It is essential that non-commercial provision be approached in the fairest and most efficient way possible.

2.1.3   Reasons for sector-specific intervention

Accepting that regulation should take place leaves another question: why specifically regulate in any one sector such as telecommunications? The existing body of competition law in many countries governs the general conduct of companies, preventing them from exploiting market failure against the consumer interest. Most countries have competition and commerce regulators. It is important to be very clear why telecommunications might require additional layers of special regulations. The reasons why most countries regard sector-specific regulation as necessary for the telecommunications services industry are threefold.

- This industry started in most countries with a dominant operator and hence an entrenched, monopolistic supply structure. To dismantle this and replace it with competition requires ongoing, proactive regulation to nourish and preserve that competition.
- This industry has some specific market failures within it. These include monopoly, high levels of vertical integration and network externalities. These require focused remedies.
- This industry is technically extremely complex. Its regulation requires sector-specific expertise that would be unlikely to exist within a general competition regulatory body.

2.1.4   Outcomes of regulation

An industry regulator's job does not or should not include market management. This follows from the theory that markets promote allocative efficiency and economic well being. Accordingly, it is not appropriate for a regulator to seek to manage the market for specific outcomes, no matter how desirable they may seem to policy makers and other influential individuals. A belief in market forces implies that we will not know, and indeed should not know, detailed outcomes in advance. This need not preclude regulation for valid social and consumer objectives, although governments do well to heed Peter Drucker's salutary warning, that 'Government is getting more powerful, not through owning, but through regulation' [1]. Generalised outcomes may, however, serve as benchmark indicators of regulatory effectiveness, for example downward price trends, international comparisons or the dispersion of monopoly market share to new players.

2.2   Markets and market failure

2.2.1   Classic causes of market failure

The telecommunications services market in most countries contains market failure. Market failure is defined not in terms of specific outcomes, but as the failure of the market to generate efficiency in the allocation of resources. There are various causes of market failure, of which monopolistic supply is predominant in the telecommunications services industry. Monopoly gives certain companies **market power**, which is the power to exclude competitors and to act independently of the interests of their customers, competitors and suppliers.

There are seven basic causes of market failure. They are listed here in an approximate order of increasing relevance to the telecommunications services industry.

- Non-excludable goods and services.
- Inefficient exclusion.
- Missing markets.
- Information asymmetry.
- Limited common property resources.
- Externalities.
- Monopolistic supply.

2.2.1.1   Non-excludable goods and services

A **non-excludable** good or service is one where it is impossible to exclude from benefit someone who does not pay for it. Well-known examples include street lighting, national defence, health care[1] and public radio broadcasts. There can be no market and hence no *economic* balancing of supply and demand for a non-excludable benefit, as it is impossible (short of taxation) to compel anyone to pay. Broadcasters typically seek reward from taxation or the adjacent revenue stream of advertising. Digital broadcasting with encryption revolutionises the economics of the broadcast entertainment industry, as it makes the service excludable. This allows operators to charge flat-rate subscriptions for access and metered (pay-per-view) prices for individual programmes.

2.2.1.2   Inefficient exclusion

Inefficient exclusion takes place whenever a supplier with excess capacity has prices well above marginal costs. An example is admission charging for museums or public parks, where prices may be set at a high level to recover large fixed costs. The result is sub-optimal because a deterred user's utility would have exceeded any marginal costs incurred. Lightly used transportation and communication networks have low marginal costs, so off-peak rates encourage customers who would not have paid the marginal cost of peak load capacity. This garners useful incremental revenue and profit for operators. Operators need to be sensitive to the cross-elasticity of demand between peak and off-peak products, since there may be scope for substitution.

2.2.1.3   Missing markets

The phenomenon of a **missing market** arises when some good or service should be traded to obtain efficient allocation, but this does not happen. Sometimes, a service is not available despite manifest demand, for example insurance against acts of war. Where there is a missing market, suppliers or others with power to allocate may do so by preference, as when a shopkeeper holds an 'under the counter' stock for favoured customers in times of shortage. Telecommunications may be suffering from missing markets for coveted numbers and addresses (these are memorable patterns such as 654321 or abc.com) and wireless spectrum.

2.2.1.4   Information asymmetry

Information asymmetry occurs whenever one party to a transaction is considerably better informed than the other about market prices, or the consequences of purchase. The concealment of defects in the selling of houses and cars is a well-known example. Doctors and lawyers have some control over the demand for their services because customers rely on their advice to know what they need. Domestic customers of telecommunications services are often at a disadvantage when buying products or choosing tariffs they do not fully understand, although business customers may make expert purchases. There is economic inefficiency if a consumer pays too much, or buys a product that does not best suit his or her needs.

2.2.1.5   Common property resources

Nature's bounty to mankind embraces a number of exhaustible but in principle non-excludable **common property resources**. Without regulation these scarce resources would be open to all. They include fish stocks, grazing rights and road space in congested locations. Because the value of the resource consumed may not be taken properly into account by a market pricing system, there is a tendency to over-exploitation. A level of consumption that depletes total utility is value-destructive.

A value-destructive level of consumption takes place if the individual user's self-benefit is greater than the personal costs borne by him or her, yet is less than the total costs imposed by that usage on the community as a whole[2]. Scarce common resources within the telecommunications arena include **radio spectrum** and **numbering and addressing capacity**.

2.2.1.6 Externalities

Externalities exist wherever an economic action imposes costs (for negative externalities) or confers benefits (in the case of positive externalities) on third parties who do not participate in the transaction. Externalities are not taken into account by the price mechanism of the market and so lead to economic inefficiency. Common examples of **negative externality** are pollution, radio interference and road congestion. There is a **positive externality** in transport and communications networks, the **network externality,** whereby all users benefit when a new destination is added to the network. Ideally for economic efficiency, consumption should be stimulated beyond the equilibrium point when there are positive externalities and restrained in the face of negative externalities. This may be achieved by internalising the externality, for example with the trading of priced permits to pollute. The alternative is to impose measures such as taxation, subsidy, cross-subsidy, price control or the imposition of quotas.

The telecommunications industry, being based on interconnected networks, contains many externalities. The positive externality of a new connection justifies subsidy of access line rentals. All networks in a competitive market must interconnect with one another, since any derogation of the any-to-any interconnection facility between terminals would inflict grave negative utility on users. A network operator that varied its technical standards of operation or of interoperation with other networks could impose high costs on others. The availability of directory services increases the value of telecommunications services facilities and stimulates usage. The utility to its recipient of an incoming call paid for by the caller is strictly speaking an externality, although it is rarely treated as one.

2.2.1.7 Monopolistic supply

A **monopoly** supplier has substantial power to set prices that diverge significantly from costs, and this militates against economic efficiency in many ways. The theoretical deadweight loss of the profit-maximising monopoly was discussed in Chapter 1. More fundamentally, a monopolist does not face the threats that would force a player in a competitive market to provide a cost-efficient service. A monopolist can be less than keen to develop best practice and innovative products, whether to bolster short-run profits, to preserve an easy life or to suit the whims of powerful individuals. A monopoly can, however, exert its market power benignly, in effect and at its discretion opting for a degree of self-regulation.

Monopoly power may crop up unexpectedly wherever there is a bottleneck point or resource, similar to that enjoyed by the owner of a strategic bridge across a river. All network operators have a monopoly for call termination, since for any one particular destination there can be no choice of terminating network supplier. Some operators exploit this market power by charging far more for terminating interconnecting calls than they would for their own subscribers' calls, even though the resource cost is substantially the same. Another bottleneck point occurs in the collation, providing and dissemination of directory information. Excessive dependence on a single resource, such as an encryption algorithm or set of keys for digital broadcasting, or the software operating system used in common access terminals, may give monopoly power to their suppliers.

2.2.2   Classification of markets and market power

A market may be classified as:

- a **competitive market**, where market forces operate reasonably efficiently;
- a **non-competitive market**, where at least one supplier possesses market power;
- a partly competitive market, where competition is present but is in some way restricted, resulting in a degree of market power.

A pre-requisite to any objectively based regulation is **market definition** followed by **market analysis**. These are the principles normally followed in the application of competition law. The first identifies the markets that exist, while the second examines those markets for the existence or absence of market power. A **market** is defined by a set of customers, of suppliers and of a distinct product (or product set), such that any supplier could serve any customer. Products that are close substitutes for one another should usually be viewed as parts of the same market. For this reason, regulation should normally be technology-neutral. A **derived market** is separate from but dependent upon another, as is the case with the wholesale markets that serve a retail market. As an example, the retail supply of Internet access depends on a derived market for the supply of PSTN access. The broadband services market similarly depends on a derived market for local loop access. Any given operator can have market power, or dominance, in one market but not in another in which it participates. The telecommunications services industry contains many markets, including:

- basic local network access;
- call minutes (local, long distance and international);
- mobile telephony;
- number translation services;
- information services;
- leased line services;
- various types of apparatus supply.

A market may be effectively competitive when its derived markets are not, and vice versa. Markets can evolve, and an interesting example is the market for voice tele-phony service carried by Internet packet technology (**Voice over Internet Protocol**, VoIP). It currently serves two market segments, a limited enthusiast market and a wholesale carriage market. If (or when) it were to become a viable substitute for traditional switched voice service, it would then converge with the normal PSTN call market. Such a market transformation might make it properly subject to regulatory attention that may have been inappropriate before.

Oftel provides a list [2] of outcomes and features that would be positive indicators of a competitive market. Consumer oriented features would include the following.

- Consumers enjoy the best or near-best deal as compared with similar economies.
- A wide range of services is available.
- Consumers are satisfied with service quality.
- Prices are broadly reflective of underlying costs (i.e. there is absence of persistent monopoly profits).
- Consumers are able to access information to make effective choices, and are confident in using this information to take advantage of opportunities.

Supplier behaviour would demonstrate these characteristics.

- They compete actively on price, quality and innovation.
- Anti-competitive behaviour is absent.
- Collusion is not present.
- Service provision meets consumer needs and is efficient.

Structural indicators of a competitive marketplace might include these following descriptions.

- There are no inefficient suppliers.
- Consumers do not face barriers to switch suppliers.
- Entry barriers are limited, so that threat of entry is a competitive discipline.
- Operators with market power in related markets (through vertical or horizontal integration) have limited opportunity to lever this dominance into the market segment in question.
- Players may be found who have recently entered the market.
- The market structure displays change over time, especially by reducing concentration.

The European Commission provides a complementary and comprehensive check-list of market and company characteristics whose existence can be taken as evidence of market power [3], as follows.

- Overall size of company.
- Control of infrastructure not readily duplicated.
- Technological advantage or superiority.
- Absence of countervailing power.
- Easy or privileged access to capital markets and sources of finance.
- Economies of scale.
- Economies of scope.
- Vertical integration.
- Developed distribution and sales network.
- Absence of potential competition.

Wherever a market is classified as non-competitive (or partly competitive), it is necessary to identify the player or players possessing market power and the reasons for that market power. The simplest case of dominance occurs when a single player has market power, although joint dominance is also a possibility. This will occur where there is collusion. However, collective dominance can occur without actual or proven collusion when two or more companies are in a position to anticipate one another's behaviour and so align their actions[3].

Regulators practise **asymmetric regulation** in the telecommunications services industry, where the different market players are not treated alike. This follows because the market has established players with market power. Telecommunications regulators thus divide their markets into players with **significant market power** (SMP) and the remainder, singling out the SMP player or players for tougher regulation. In the case of national fixed wireline markets, the incumbent monopolist will very obviously be an SMP player.

2.2.3   Working definitions of market power

When telecommunications services markets were first liberalised, there was usually a very clearly dominant player, namely the incumbent and former monopolist. The concept of SMP in European regulation was then based on the simple structural measure of their market share. This allowed the application of targeted regulation to the incumbent. The threshold value for SMP was originally set in Europe at 25 per cent market share (Article 4.3 of Reference 4), although more recent thinking has been suggestive of a figure of 40 per cent [3]. These measures are not to be taken precisely, to the extent that a player would be deemed to have SMP with 40.5 per cent of a market but not with 39.5 per cent. Under such circumstances its regulatory classification as SMP might oscillate, giving it an incentive to conceal its market share by creative accounting or even by turning away customers (at least in the month before statistics are collected). A near approach by a player to a threshold should, therefore, trigger no more than a careful re-evaluation and certainly not automatic reclassification.

An elaborated measure of market concentration is the Herfindahl-Hirschmann Index (HHI) (Equation 2.1), where S_i is the fractional market share of the ith player in the market. The HHI can never exceed 10,000, its value in a total monopoly. However, this and all other structural measures are open to the criticism that a true definition of market power should be a behavioural definition.

Equation 2.1   Herfindahl-Hirschmann index of market dominance

$$HHI = \sum_i (100S_i)^2 \qquad (2.1)$$

A player with significant market power can be defined as one who can price above the competitive level for a non-transitory period without losing market share. The competitive level is the long-run incremental cost[4] of that industry for that product or service. Even a monopolist may not possess market power in this sense if entry barriers are low, that is, if the monopoly is contestable. In such circumstances it would not be valid to classify it as having SMP. For example, a supplier may take an 80 per cent or even 100 per cent market share for a novel piece of software, yet the monopoly may be potentially transitory, being vulnerable to another supplier's improved product. The Lerner index, the proportionate deviation of a price above the marginal cost of production, is one tool for investigating the existence of market power, as is price elasticity of demand, since a supplier with inelastic demand can always increase revenue by raising prices.

With markets becoming more nearly or prospectively competitive, the definition in European regulation is shifting from structural measures towards the competition law concept of **dominance**, as amplified by European case law. This is defined (and elaborated by considerable discussion) in Reference 3 as follows:

> An undertaking shall be deemed to have significant market power if, either individually or jointly with others, it enjoys a position of economic strength affording it the power to behave to an appreciable extent independently of competitors, customers and ultimately consumers.

2.3 Abuse of market power

2.3.1 Overview

Companies possessing market power can exert that power in ways that prevent others from entering the market. These are known as **abuses of market power** or unfair competitive practices, since their effect is to stop the consumer from enjoying a better or cheaper service than he or she could otherwise have obtained from a competitor. A key role of the telecommunications regulator is to identify market abuse, and to intervene in the market to prevent it. It is, of course, no part of the regulator's role to prevent a company with or without market power from succeeding and winning customers by *fair* means, where it is able to provide better and more efficient services. The types of anti-competitive practice against which telecommunications regulators typically need to act are as follows.

- Predatory pricing.
- Margin squeeze.
- Discriminatory pricing.
- Product bundling (linked sales).
- Exclusive dealing arrangements.
- Cross-subsidy.
- Control of essential intellectual property.
- Information sharing.

2.3.2 Predatory pricing

Predatory pricing occurs when a company deliberately sells a product or service below its true cost with the aim of excluding competition and raising prices subsequently to offset initial losses. In the most extreme form of predatory pricing, the seller may buy market share by incurring a loss on each sale. Large companies may practise predatory pricing more subtly without direct loss, by pricing to recover a less than proportionate amount of fixed cost, in effect cross-subsidising sales from profits on other services. The result is to place pressure on competitors either to match the lower price or to leave the market. The predator, being strong in financial resource, can sustain losses longer than the competitor, who must eventually leave the market or go out of business altogether. At this point, the predator can reassert its powerful position and use its market power to raise prices and recoup earlier losses.

Predatory pricing in regulated telecommunications services markets is not necessarily a very serious problem, for the simple reason that its existence is often (though not always) plain to see. There are two classic tests for predatory pricing, the stand-alone cost test, and the incremental cost test. The **stand-alone cost** (SAC) of providing a service is the cost of supply that would be incurred by a supplier who supplied only that good or service, and pricing below this is an indication of predatory intent. The incremental cost of providing the service is the difference between the total costs of the supplier with and without the service in question, and again revenue below this is evidence of predatory pricing. Neither of these tests is completely satisfactory, however, if applied individually to single services in the presence of related services, since some cross-subsidy may escape these tests.

2.3.3   Margin squeeze

Margin squeeze may occur whenever a supplier with market power provides a wholesale input to a company with which it also competes in the provision of the corresponding end-user retail service. Wholesale transport of long-distance calls by incumbent telecommunication operators is a common example. It is potentially a more serious concern for the regulator than predatory pricing, on account of its relative difficulty of detection. The principle of margin squeeze is illustrated in Figure 2.1.

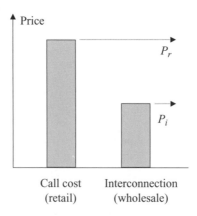

Figure 2.1   Simple illustration of margin squeeze in call termination.

It shows an SMP network operator who terminates a call for another operator at the wholesale interconnect price P_i (this would normally be in units of pence per minute), and connects calls for its own customers at the retail price P_r. The margin, $P_r - P_i$, is the amount available to the competitor for its costs and profit. Margin squeeze takes place when the SMP operator raises the wholesale price and reduces the retail price. Together, these may be neutral with respect to the SMP operator's overall profit, yet considerably weaken the competitor's business.

A simple test for margin squeeze is based on the avoided cost to the SMP operator of not supplying the downstream retail product. So long as the avoided cost does not exceed the margin ($P_r - P_i$), then there is no margin squeeze. This simple test needs some elaboration as shown in Equation 2.2 to be realistic, however, since the SMP operator's vertical integration gives it an advantage of a cheaper internal transfer when selling a wholesale service to itself than to an external party. This is an economy of scope that it ought to be allowed to pass onto customers. In Equation 2.2, M is the margin, P_r and P_i have the same meanings as before and in Figure 2.1, C_v is the avoided cost of not supplying the retail output, C_e is the cost of supplying the intermediate output to an external party, and C_i is the equivalent cost of internal supply. The mathematics may be straightforward, but obtaining the correct values is far from simple.

Equation 2.2   Test for absence of margin squeeze

$$M = P_r - P_i \geq C_v - (C_e - C_i) \tag{2.2}$$

2.3.4   Discrimination

Discrimination takes place whenever a supplier sells the same product at different prices in different markets or to different customers, and where there is no objective economic justification for that difference. Individual negotiation of prices and discounts is a common feature of the competitive sales environment, and may represent both the workings of the process by which the market determines the price of a good or service, and also an element of discriminatory pricing. SMP players are typically prevented by regulation from discriminatory pricing, since otherwise it would be too ready a vehicle for predatory pricing in desirable customer accounts. Typical national monopoly operators find that as much as half their profit may come from 10 or 5 per cent of their customers, so they would have a ready motivation to use predatory prices to retain these accounts and stop them migrating to competitors. In effect, it would be cross-subsidising these accounts from monopoly profits on smaller accounts and from customers with less propensity to change supplier.

2.3.5   Bundling or linkage of sales

Bundling or **linkage of sales** occurs whenever products and services are sold in combinations chosen by the supplier. Some groupings of products are unexceptionable, when the combined items are so integral one with the other that no one could, or would want to, take them separately. An example is basic exchange line service with its maintenance. However, bundling provides an opportunity for market abuse, were the dominant supplier to group products together in a way that forced customers wanting something obtainable only from it (say, the basic exchange line) also to purchase other products (such as long distance service) that were neither wanted nor needed. Pricing can have the effect of linkage when a product is offered at a high price in isolation but at a more realistic or desirable price when taken as part of a combined offering.

2.3.6   Exclusive dealing arrangements

Exclusive dealing arrangements of various kinds can be used to unfairly disadvantage competitors. Many companies negotiate exclusive arrangements with suppliers, usually obtaining discounts in return for granting a sole supplier role and commitment to purchase certain volumes, and these arrangements have economic logic and are not necessarily anti-competitive. They become unfair if a company with large purchasing power exerts this power to persuade a supplier to agree *not* to supply certain items to its competitors.

2.3.7   Cross-subsidy

A large company with horizontal or vertical integration may be profitable as a whole while having less profitable and even loss-making products in its portfolio. Where this takes place, **cross-subsidy** is said to exist, where the profits on some items pay for the costs of others. Cross-subsidy is not sustainable under conditions of true competition, since it would lead to capture of loss-making business from competitors while simultaneously ceding the more profitable lines to them. However, a monopolist or SMP player may use its price-setting power to produce ongoing cross-subsidy. Although we have seen that some forms of cross-subsidy may be benign when they address externalities, others provide an opportunity for predatory pricing and the exclusion of competition. A typical example might be the cross-subsidy by an SMP operator from its monopoly network operation to its competitive business of apparatus supply.

Tests for cross-subsidy are complex. The stand-alone cost and incremental cost tests (as described above for predatory pricing) are insufficient to prove the absence of cross-subsidy. It is necessary to ensure that the incremental cost of a product at a particular price is at least equalled by its net incremental revenue after subtracting the lost revenue effects arising from substitution of this product for others at their prevailing prices[5]. Regulators typically require SMP players to divide their business into parts and provide separate accounts for each under laid-down standards of accounting practice, and this allows at least gross cross-subsidies to be detected. Minor cross-subsidy may in practice be less damaging than the cost of policing for it.

2.3.8   Intellectual property and information sharing

The way an SMP operator uses and shares, or fails to share, intellectual property provides opportunities for anti-competitive practice. At the very simplest, it must disclose enough to competitors to allow them to interconnect with its network, and this information will include details about the location of exchanges and cable joint boxes, the types of switching and transmission equipment in use and the signalling systems and protocols employed. This amount of disclosure may prove uncomfortable in cultures where telephone exchanges were (and are) possibly classified as national security sites, so the most detailed disclosures may be restricted only to other operators, that is those having the need to know. Network information transmitted in either direction between interconnecting network operators may have to be accepted and treated carefully as confidential information. Where intellectual property has economic value, for example where the former monopoly operator owns the rights to a national variant of the inter-exchange switching system, the SMP operator must release it at a fair, cost-related, value and cannot be allowed to use its undoubted market power to set discouragement prices.

If an operator holds patents necessary for interconnection with its network or the operation of a similar network, then failure to license them to other players (or to ask a price so high that it was equivalent to refusal to supply) would exclude them from the market.

A former monopoly operator will, by the nature of its operation, obtain a great deal of knowledge about its customers and their movements. Sharing this, especially with its sales forces, is frequently considered an abuse of the dominant position. For example, suppose a customer orders a private circuit to a competitor's telephone exchange. The provisioning department at the SMP operator that handles this may be forced to practise information separation, being prohibited from passing this knowledge on to the long-distance telephony services salespeople. They might then attempt persuasion of the customer, or pass on a tip that the customer may be looking for a PBX. Where the regulator has reason to believe that certain classes of customer may be unaware of competitive sources of supply, it may require the SMP operator's people and literature to make it clear that other suppliers exist.

2.4   The rules of regulation

2.4.1   Overview

This section provides an overview of the rules that are typically brought to bear on the telecommunications services market. Most of what follows refers to the regulation of a company classed as having SMP, although some regulation is appropriate for non-dominant players also. This section must be regarded only as a guide, however, as the detail varies over space and time. Different countries have different formulations, and all change and adapt regulation to new challenges and situations. The rules described here are based loosely on those applying to BT at the end of 1995 [6]. Regulatory rules may be known variously as 'orders', 'regulations', 'licence conditions' or similar terms in different countries, though the word 'rule' is used here.

It is helpful to analyse the content of regulatory provisions under four headings representing different categories of regulatory activity, as illustrated in Figure 2.2. Regulation implements a national telecommunications policy through administrative, social and consumer goals. Many, though not all, consumer objectives are served by

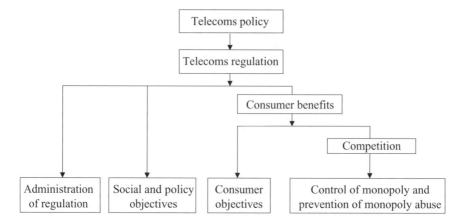

Figure 2.2   Arenas of regulatory action.

the creation and nourishment of competition. A raft of controls designed mainly to prevent monopolistic abuse of power encourages and enables competition. These act as a proxy for the missing competitive discipline in the incompletely competitive market. Real-life measures do not necessarily, of course, map exclusively onto one of the category headings in Figure 2.2, and may overlap.

2.4.2   Administrative rules

Administrative regulation defines the businesses in which a company may engage, and any other rules necessary to allow the regulator to perform its role. When the company is a former monopolist, as was the case with BT, some role definitions may be a *de facto* statement of things it did before, for example international and maritime services. Rules of a purely administrative nature may include:

- a statement of the basis on which the company must provide its financial accounts;
- a requirement to pay a fee to fund the activity of regulation;
- a requirement to provide information of any type when so requested by the regulator (normally subject to a condition of practicability and restricted only to information that the company can reasonably be expected to have in order to run its business);
- a consent of release from obligations under *force majeure*, impracticability, or where the obligation might otherwise operate in favour of persons or companies of proven bad record.

A company's accounts might normally be prepared in accordance with the standard statements of accounting practice in the country concerned, although the regulator may need to supplement these, for example, where it requires summary accounts more often than annually, or where it wishes to be more prescriptive about the mechanisms for valuing assets or reporting internal transfers between component businesses.

2.4.3   Service obligations and social regulation

Social objective rules oblige a company to provide certain classes of service, and remove its commercial freedom not to provide them or to change their nature or prices without consultation. There may be a requirement to provide these on a non-commercial basis at 'affordable prices', either to all or to disadvantaged elements of society. Where the cost of meeting an obligation is significant, some mechanism may be prescribed for the recovery of the costs from the whole industry. Typical obligations are as follows.

- An obligation to provide universal basic service across the whole territory, including in rural areas. This is known as a **universal service obligation** (USO). This obligation need not extend to all the services that may be variously offered by the operator, but will refer to a defined set including simple voice telephone service.
- A requirement for geographical price averaging, that is to provide uniform prices to all customers across the whole territory. Provision is usually made, however, for higher charges for installation work requiring more than a prescribed cost or number of hours' labour.
- An obligation to provide public telephones. While individual installations may be removed, this can only be done where the revenue of the call boxes falls below determined limits, where means of alternative provision have been explored, and after due consultations with local bodies.
- A requirement to provide a public emergency call service.
- A requirement to give priority and assistance to calls made by emergency organisations such as the fire, police, ambulance, coastal and mountain rescue services.
- A requirement to help emergency organisations and other public bodies make plans for the handling of local and national emergencies.
- A requirement to provide a continuous, priority maintenance service to customers of types determined by the regulator, and adequate service to others.
- A requirement to provide a reduced price service to low usage customers. This is a regulated service; were the company to exceed its obligations or make offers to customers beyond the terms of this obligation, it might fall guilty of discriminatory pricing.
- A requirement to provide apparatus and services to assist users with disabilities and cognitive impairments, and to consult bodies representing the interests of such users. In practice, such rules are apt to become very detailed where they relate to specific requirements, such as telephones and call box facilities for those with impaired hearing, and directory facilities for sight-impaired users.

Universal service obligations bring benefits as well as costs to the company providing them. They give their provider a ubiquitous brand position that draws customers towards the operator they know. Their costs may not be nearly as high as is sometimes believed. A study in 1997 and reported in Reference 6 speculatively suggested that this brought 223,000 moving customers per annum to BT who might otherwise have contracted with a competitor, thus offsetting costs estimated in the same study to be 65 million – 85 million pounds. This was only about 1 per cent of BT network revenue. The Federal Communications Commission (FCC) provided an estimate of 12 billion dollars per annum for the annual subsidies for universal service in the USA in 1997.

2.4.4   Consumer regulation

Consumer benefit is expected mainly to flow from competition in the market place, so regulations relating exclusively to consumer affairs tend to be relatively minor. They may include the following.

- An obligation to provide a minimum set of leased line services.
- An obligation to provide a simple maintenance package for single exchange lines at a regulated price. Premium or high-priority maintenance services are not necessarily affected by this rule.
- A right for a customer to obtain installation of a second line at the same premises as an existing line while paying for the installation by instalments.
- A requirement to ensure that metering and billing is accurate, and to give the regulator reasonable access and means of verifying this.
- A requirement to provide itemised bills. This is in practice a dated rule, given that most operators now offer this and an operator not so doing would suffer competitive disadvantage. Back in the 1980s, this type of rule provided an incentive to bring about necessary technical development of exchange and billing systems.
- A requirement not to offer certain, listed types of service until a code of practice has been agreed with the regulator. These services, known as 'controlled services' or 'special services', may include Premium Rate services, 'Adult' services, chat lines and live conversation services.
- A requirement to provide a facility allowing customers to opt into (or opt out of) access to special or controlled services, and to provide itemised billing showing clearly the amount of spending on such services.
- A requirement to have a code of practice for the handling of confidential customer information.
- A requirement to have a process for handling customer complaints, and to review it periodically with the regulator.
- A requirement to have a process for arbitrating customer disputes, and to review it periodically with the regulator.
- A requirement to consider and respond to matters raised by various representative and advisory bodies. The bodies to which attention must be paid may be prescribed by the regulator or by other laws.
- A requirement to take reasonable precautions to ensure the privacy of customer information (for example, of unencrypted radio), and to provide the customer with control over the divulging of information (for example, of calling line identity).

2.4.5 Regulation to promote competition

A great deal of the regulation of an SMP operator relates to the control of monopoly and to the prevention of abuse of monopoly power. Some provisions simulate the disciplines that would otherwise have been exerted by market competition, and others are straightforward prohibitions of unfair, anti-competitive practice. The following headings do not have any significant order of importance.

2.4.5.1 Price control

One of the regulator's most powerful tools for inserting the missing discipline of a market is control of prices. This complex topic is described separately in Chapter 4.

2.4.5.2 Network integrity regulation

Given the fundamental importance of telecommunications in the modern economy, regulators may lay down guidelines relating to the availability, resilience and physical security of telecommunications networks. These will address, among other things, special protection of emergency services and the ongoing provision of service to government, the emergency services and general consumers in the presence of:

- catastrophic network failures;
- extreme weather and natural disasters;
- wartime threat and terrorist attack;
- traffic surges and focused overloads;
- incorrect operation of terminal equipment and interconnected networks;
- malicious network attack and denial-of-service attack.

Operators may be required to respond directly to the requirements of government departments and local administrations in times of national emergency, and to liaise with them regarding advance planning for emergencies and service restoration in case of emergency.

2.4.5.3 Quality of service regulation

Regulators tend to avoid, where practicable, the detailed prescription of the quality of consumer and business services, seeing it as the prerogative of customers in a competitive market to decide the qualities of service for which they wish to pay. New entrants have frequently bought network equipment from the same suppliers as former incumbents, adopting the interconnection standards developed during monopoly years. As a result, call quality has been a relatively minor issue. Whether this will continue to be the case with technological innovation remains to be seen.

Quality-of-service obligations normally take a subtle role in present-day regulation. They are employed to ensure that companies cannot use quality degradation as a way of satisfying price controls, nor as a tool of discrimination between customers. A company may be required to document its quality of service, and permit the regulator reasonable access to its premises, processes and records to investigate any alleged or suspected acts of discrimination or reduction of quality. Regulators may publish comparative statistics, for example about repair times, call failure and network availability, to better equip customers to judge the relative worth of supplier offerings.

Legislation on specific issues such as privacy of radio communications [7] or security of encrypted messages may bring about the regulatory setting of standards.

2.4.5.4   Interconnection

Interconnection is fundamental to the operation of a competitive market, and ensures the necessary facility of any-to-any connection between compatible devices in national and international networks. Players with market power are normally compelled to interconnect with other operators. It can normally be assumed that non-SMP players will choose in their own interest to connect with other operators. The requirement on an SMP operator to interconnect prevents a possible monopoly abuse, because only the former monopolist could maintain a viable business while refusing to interconnect. Rules normally oblige an SMP operator to provide a number of distinct facilities under the umbrella of interconnection:

- to terminate calls destined for its own subscribers;
- to connect calls from one network operator to another ('transit' carriage);
- to allow other operators' customers access to those operators' networks through its network and access lines ('indirect access');
- to permit resale of its basic services, for example to allow others to provide a voice telephony service over circuits leased from it.

Regulatory price control of interconnect services is critical, since an SMP player would in general have no incentive to minimise its rates. Price control can be relaxed, however, where use of the SMP operator's service is avoidable. In that case, customers would have effective choice and the market would be competitive. This situation applies in practice to the long distance transit market in many countries.

Be in no doubt that the requirement to interconnect imposes a cost on the industry and hence on consumers. This arises through the technical costs of interconnection and of interconnection accounting, and also because interconnecting calls may take less efficient routeings than would have been the case within a monopoly network. Regulators make the judgement that such costs are more than outweighed by the economic benefits resulting from competition.

2.4.5.5   Technical interfaces: essential interfaces

Where networks are to interconnect, then it is essential that they do so compatibly, and regulators retain the right to mandate the use of certain technical interfaces. In many countries, this has proved more of a reserve rule than an active rule, since operators in many markets agree their interfaces without intervention. There may be *de facto* adoption of the SMP player's interfaces in the absence of a reason to do anything else. Regulations may be framed to give bias towards open international interconnect standards, requiring the companies to move to open standards, and justify each instance where one requires a non-open standard for access to its systems. Operators with market power will be required to consult extensively and give plenty of notice (in the UK, six months) before making any change to a technical interface by which its systems are accessed.

Regulators may be expected to work closely with standards bodies such as the **International Telecommunications Union** (ITU) and the **European Telecommunications Standards Institute** (ETSI), asking them to help prepare standards where necessary to ensure harmonious interoperability between competing networks and facilities.

2.4.5.6 Pre-notification of prices

The pre-publication of price changes a specified number of days in advance (28 in the UK) gives competitors an opportunity to prepare, react and, if necessary, object. For wholesale prices, an SMP operator may be obliged to publish a formal **interconnect offer** detailing its wholesale rates. This enables other operators to know the interconnect prices they will have to pay and so construct business plans.

2.4.5.7 Numbering and number portability

Operators are bound to abide by national numbering conventions, which include the procedures established by the regulator for applying for numbering capacity and providing periodic returns about its utilisation. A requirement to provide number portability, that is the facility for a consumer to change service supplier while retaining his or her number, is usually mandated.

2.4.5.8 Provision of directory information

The provision of directory information by any means, whether by printed material, machine readable material, voice enquiry service or on-line based enquiry service, represents a bottleneck service through which a dominant operator could disadvantage other operators. It is thus required to do two things. The first is to include in its various directory products the numbers of customers of other operators, to the extent that the other operators furnish this information in a reasonably usable form. The second is to make available directory information about its own and other operators' customers to another operator who requests it for the purpose of providing directory services and for connecting calls.

2.4.5.9 Notification of mergers, acquisitions and joint ventures

Where an operator proposes any form of merger, partnership, association or joint venture with another in pursuit of a regulated activity, then it must notify the regulator of any such arrangement a specified number of days in advance (in the UK, 30). It is then open to the regulator to decide whether such action might have the effect of creating a monopoly, or of materially changing the significant market power status of a player in the market. If a company has associate companies over which it has corporate control, then these must accept regulatory obligations and abide by prohibitions as for the parent.

2.4.5.10   Accounting separation

Accounting separation requires the operator with market power to divide its operations into discrete businesses, for example its network, mobile network, apparatus supply, systems management and consultancy, value-added services and equipment manufacture, and to submit separate accounts for each. It must prepare these accounts under accounting conventions and at intervals (not more than quarterly in the UK) laid down by the regulator. This is the principal mechanism by which the regulator ensures that there is no cross-subsidy. There may be specific prohibition of cross-subsidy, and a requirement to make known material transfers between component businesses. The regulator usually retains discretion to determine whether a particular cross-subsidy is unfair or not, bearing in mind that it may have taken place to satisfy a social obligation. Where there is an equipment-manufacturing subsidiary, the regulator may oblige the company to adopt an open tendering approach to apparatus procurement.

2.4.5.11   Non-discrimination

An operator with market power must publish its prices and apply them in an even-handed way to all customers. This applies to retail, wholesale and interconnection prices. The operator may be required to show that its wholesale prices are even-handed with respect to the transfer rate used for the supply of the same services to its own retail business. Where the operator has systems and value-added services businesses, it must ensure that other players in this market may obtain its basic services at the same prices it charges internally. Quality of service may not be varied in a discriminatory way, so placing competitors' businesses at a disadvantage. Quantity discounts may be allowed and may not be classed as discriminatory, as long as they are available equally to any customer who meets the published qualifications for them.

2.4.5.12   Prohibition of linked sales

Linked sales occur when a company makes the purchase of one good or service dependent on the purchase of another, and these are prohibited excepting where joint purchase is essential. Pure discounts for quantity are not prohibited, however. The regulator may provide guidelines for acceptable differential pricing of linked purchases.

2.4.5.13   Prohibition of exclusive dealing arrangements

The prohibition of exclusive dealing arrangements is directed at stopping a company from negotiating deals with suppliers that have the effect of making its purchase conditional on the supplier not supplying similar equipment to other companies or persons. This should not, however, prevent it from obtaining exclusivity on minor features (such as the logo on its apparatus), nor on retaining intellectual property rights where it made an input to design and development.

2.4.5.14   Fair access to intellectual property

A dominant operator must do all it can to ensure that other operators, resellers and customers can obtain at a fair and reasonable price any intellectual property necessary to connect with its network. The regulator may have powers to determine the rules on which such rights are granted to others, respecting that it cannot enforce an action that would infringe a third party's rights or cause the company to forfeit the licence under which it holds the intellectual property rights.

Where a company is the holder of a patent, and failure to license the patent to others results in a barrier to entering the market, it may be obliged to license others to use it. It may not set a price so high that it has the same effect as refusal to supply. It should have the right to license it at fair and reasonable cost. It should have the right to recompense if, after the granting of a licence, onward sales of products and systems are made to further companies.

2.4.5.15   Testing requirements for interconnection and connection of apparatus

Testing requirements are an area where a company might practise abuse of a dominant position. An incumbent operator is entitled to test any network or apparatus proposed for connection to its network, for conformance to standards for safety, interoperability and network integrity. However, the testing requirements must be objectively justified, and the company may not insert features of its own devising in addition to agreed standards unless the regulator shall so consent.

2.4.6   *Controls on operators without market power*

While companies without market power may expect to experience less regulation than their dominant competitors, some regulation is essential to ensure that the market remains fairly competitive or indeed operates at all. Typical regulation bearing on non-dominant players might include:

- essential technical interfaces;
- network integrity;
- conformance with the national numbering scheme;
- provision of emergency services;
- support of number portability;
- accuracy of billing;
- pre-notification of joint ventures, mergers and acquisitions;
- requirements to furnish various statistical returns, such as market share data, usage data and numbering utilisation;
- the entitlement of a consumer to a contract, and minimum requirements for what the contract should specify;
- payment of a fee to the regulator.

2.5   Notes

1   People gain utility from other peoples' healthcare because they are less vulnerable to infection when living in a healthier population.

2   This phenomenon is demonstrated by the 'commons problem' or 'tragedy of the common'. The individual who grazes one extra sheep gains a unit of personal benefit that is less than the total yield loss inflicted on all the others' under-nourished sheep.

3   Collective dominance is explored in Reference 3, Sections 3.1.2 and 3.1.3.

4   The concept of a long run incremental cost (LRIC) will be introduced in detail in Section 4.4.4 dealing with price control. For the present, it should be regarded as a marginal cost that includes essential fixed costs or in other words takes a long-run view of variable costs.

5   Readers requiring a more detailed discussion are referred to Reference 5.

2.6   References

1   *Financial Times*, 27th April 1999 (quoted in Reference 8)

2   'Oftel's 2000/01 effective competition review of dial-up Internet access', Oftel, 30th July 2001, paragraph 2.2

3   'Draft guidelines on market analysis and the definition of significant market power', EC Working Document COM (2001) 175, March 2001. The European Commission, Brussels

4   Directive 97/33/EC of the European Parliament and of the Council of 30th June 1997 on interconnection in telecommunications with regard to ensuring universal service and interoperability through application of the principles of Open Network Provision (ONP). The European Parliament, Burssels, 1997

5   WHEATLEY, J. J.: 'World Telecommunications Economics' (IEE Books, London, 1999)

6   BT Operating Licence, November 1995 (HMSO, London)

7   'European convention for the protection of human rights and fundamental freedoms', Article 8. The Council of Europe, Rome, 1950

8   HALL, C., SCOTT, C., and HOOD, C.: 'Telecommunications regulation – Culture, chaos and interdependence inside the regulatory process' (Routledge Advances in Business and Management Studies, London, 2001)

Chapter 3

The framework for regulation

3.1 Legal frameworks

3.1.1 Regulation and the law

This section provides a brief overview of the legislative background to the regulation of telecommunications. Readers requiring more detail about legislation and case law should consult one of the legal guides and textbooks, such as References 1 or 2. This is a rapidly moving field. For example, the 1996 Telecommunications Act of the USA revised the previous legal framework drawn from the 1934 Act, while the European Union revised its **Open Network Provision** (ONP) framework directives from around 1990 in the period 1997–1998, and has done so again in 2002. This new European framework, which comes into effect in July 2003, is summarised in some detail in the Appendix.

Figure 3.1 provides a schematic overview of the process of telecommunications regulation.

Laws passed by governments define and enable the process of regulation. These laws appoint regulatory bodies and set their terms of reference. They define their objectives, operating principles and accountability, and confer on them powers to take actions. Laws that bear specifically on the telecommunications services industry include[1]:

- sector-specific telecommunications law;
- competition and fair trading law;
- other relevant law, for example concerning privacy, consumer protection and rights to build infrastructure.

Laws are created by national legislatures. Regional law-making bodies, such as the **European Union** (EU), or world law-making bodies such as the **World Trade Organisation** (WTO), may also enact relevant provisions. These laws, directives or agreements take force in member and signatory countries when they are **transposed**

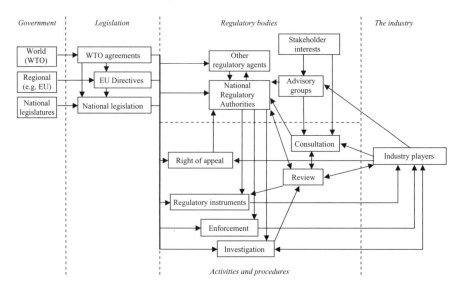

Figure 3.1   Overview of telecommunications regulation.

into national law. Countries do this in accordance with treaty obligations, for example under the European Union's founding Treaty of Rome of 1957.

The function of regulation may be distributed between two or more separate bodies, for example:

- a formally designated national regulatory authority;
- government departments;
- historically established bodies, such as standards organisations and radio-communications agencies.

Countries may further sub-divide their regulatory function, for example into federal and state regulators.

Regulators have to take account of stakeholder interests, which may be prescribed by law. These include consumer bodies, industry bodies, local authorities and planning bodies. Specialist organisations also have interests where appropriate, for example those having care for environmental conservation, national parks or areas of out-standing natural beauty. Standing advisory groups may be formed to gather necessary expertise and represent stakeholder interests.

Regulation is accomplished by means of instruments and processes. The main instruments of regulation are described in Section 3.3 and include authorisations, rules, determinations, consents and guidelines. The instruments of regulation are backed by powers of enforcement and of investigation. They are subject to periodic or *ad hoc* review, and in many cases the law requires a process of consultation before draft instruments take effect. Regulatory actions are subject to a right of appeal, either to bodies specifically constituted for the purpose or, ultimately, to the courts.

3.1.2 Sector-specific regulation

Most countries have recognised that some sector-specific regulation is appropriate for the telecommunications services industry, for the reasons given in Section 2.1.3. They have, accordingly, established by primary legislation **National Regulatory Authorities** (NRAs) for their telecommunications sectors, of which a selection is listed in Table 3.1.

Table 3.1 A selection of national telecommunications regulatory authorities (source: Oftel).

Country	NRA title	Acronym	Web site
Australia	Australian Competition and Consumer Commission	ACCC	www.accc.gov.au
Brazil	Agência Nacional de Telecomunicações	ANATEL	www.anatal.gov.br
Canada	Canadian Radio and TV Commission	CRTC	www.crtc.gc.ca
France	Autorité de Régulation des Télécommunications	ART	www.art-telecom.fr
Germany	Regulierungsbehörde für Telekommunikation und Post	RegTP	www.regtp.de
Hong Kong	Office of the Telecommunications Authority	OFTA	www.ofta.gov.hk
India	Telecommunications Regulatory Affairs India	TRAI	
Ireland, Republic of	Office of the Director of Telecommunications Regulation	ODTR	www.odtr.ie
Italy	L'Autorità per le Garanzie nelle Communicazioni	AGCOM	www.agcom.it
Malaysia	Malaysian Communications and Multimedia Commission		www.cmc.gov.my
Netherlands, The	Onafhankelijke Post en Telecommunicatie Autoriteit	OPTA	www.opta.nl
Singapore	Info-Communications Development Authority	IDA	www.ida.gov.sg
South Africa	South African Telecommunications Regulatory Authority	SATRA	www.satra.org.za
Spain	Comisión del Mercado de las Telecomunicaciones	CMT	www.cmt.es
Sweden	National Posts & Telecommunications Agency	PTS	www.pts.se
UK	Office of Telecommunications	Oftel	www.oftel.gov.uk
USA	Federal Communications Commission	FCC	www.fcc.gov

Regulatory bodies are reorganised from time to time, so for example the UK regulator Oftel is in process of being merged with media and radio regulation bodies to form a combined Office of Communications (Ofcom).

The mission of a National Regulatory Authority is likely to include the following elements.

- The creation, nourishing and maintenance of a competitive market for telecommunications services.
- Ensuring that the telecommunications services market meets all national and international needs.
- The promotion of the interests of consumers.

Oftel in the UK thus states [3] its overarching aim as 'to secure the best deal for the consumer', and its goals as effective competition benefiting consumers, well-informed consumers, adequately protected consumers and prevention of anti-competitive practice.

National regulatory authorities may attract subsidiary roles such as the giving of specialist advice to government departments, and the representing of the government at international meetings. They may acquire duties that could be fulfilled elsewhere but are included on a basis of convenience, in other words because 'someone has to do them'. These may include the administration of the national numbering system or the management of a number portability database. The handling of consumer complaints is a suitable adjacent function that may be handled by a regulator, as it is complementary to the regulator's competition objectives and is compatible with the contacts and expertise it will develop.

The use of the term 'regulator' or 'regulatory body' in this book refers to the holder or holders of the regulatory function. This will normally mean the designated national regulatory authority. However, and as mentioned before, the 'regulator' may be a distributed function and the NRA not the only player. As an example, the licensing role (authorising market entrants) in the UK is reserved by a government department, the Department of Trade and Industry (DTI) and not by the NRA, Oftel. Oftel does, however, monitor compliance and is responsible for licence (rule) amendments.

3.1.3   Ex-post and ex-ante regulation

A regulator's rulings may be examples of **ex-ante** or **ex-post** regulation depending on when the rules are applied.

An ex-ante regulation states, in advance, a rule that a company must obey. It might be:

- an obligation, something it *must* do;
- a prohibition, something it *must not* do;
- a conditional permission, saying that *if it does this*, then certain conditions will apply.

An ex-ante rule has the property of clear statement in advance, and an industry player subject to the rule can know, at least in principle, whether or not it complies without further reference to the regulator.

Ex-post rules have a different characteristic, where the general principles are laid down, for example, 'You must not indulge in anti-competitive practice,' but it is left to the regulator or the courts to determine after the event whether a breach has taken place. Some people see ex-post regulation as 'common sense' regulation, providing the freedom to secure justice without having to codify minute cases and conditions. Detractors may remark that it gives the regulator the combined functions of lawmaker, policeman, judge, jury and executioner. However, prior statement of objectives, governed by principles of limitation and to which appeal may be made in the courts, restricts ex-post regulation from being a merely discretionary exercise of power. A body of guidelines and accumulated case law assists the efficient administration of ex-post rules.

A strong framework of ex-ante regulation enables the proactive steps necessary to transform a monopoly market into a competitive market. While ex-ante rules give players a good idea where they stand, they are vulnerable to becoming dated or inappropriate in a fast-moving high technology industry. A company with smart lawyers may find that it can 'game' ex-ante rules in ways that dilute or nullify their original intent, especially when rules are complicated. Ex-post regulation, typified by various types of fair trading clause, gives regulation its power to react reasonably in unforeseen circumstances. By concentrating action on proven grievances rather than on compliance with finely detailed rules, ex-post regulation is potentially efficient regulation.

3.1.4 Relationship with competition law

Regulators typically operate under the terms of competition law as well as telecommunications law. They may have, as in the UK, concurrent powers with a Director of Fair Trading or similar-titled official. Competition law is formulated in general and ex-post terms about what constitutes anti-competitive practice, based on concepts of market analysis. This gives a regulator freedom to deal with anything that is actually or potentially an obstruction of competition. Nonetheless, to initiate action a regulator must spend time and investigative effort building a case. This will cite its reasons for believing that conduct is or will be anti-competitive. Such a case must normally rest on things that have happened, or on things whose likelihood of happening can reasonably be inferred from things that have happened. Telecommunications law is typically more specifically focused than competition law, allowing the regulator to pursue established and forward-looking ex-ante measures without facing a burden of proof of anti-competitive intent in every instance. In deciding whether to proceed under competition law or telecommunications law in a particular case, a regulator must use discretion in selecting the most efficient and appropriate legal basis.

As markets become more competitive, it is likely that less use will be made of sector-focused regulation and more of competition-based regulation. This is indicated in general by the term **light touch regulation**, which will have less detailed ex-ante prescription, less intrusion into day-to-day running, and more stress on general principles and ex-post rules. Ex-ante regulations created for liberalisation were historically necessary in order to classify the former monopolist as dominant and pursue proactive measures to prevent its abuse of market power. The UK's Draft Communications Bill, which proposes to establish the wider Office of Communications (Ofcom) in replacement of Oftel, requires Ofcom not to impose or maintain unnecessary burdens and to publish periodic statements showing how they propose to achieve this[2].

3.1.5   Principles of administration

Regulation within the European Union and elsewhere is governed by certain principles of **administrative law**, concerning the way that legal powers are exercised. These principles are as follows. Application of these principles protects regulation from political bargaining and interference.

- **Objectivity**   Each regulatory action must have a stated objective, and be consistent with the overall declared objectives of the regulatory process.
- **Reasonableness**   It must be possible to explain and justify the steps being taken.
- **Appropriateness**   Regulatory actions must address the problem being solved.
- **Proportionality**   Regulation must be proportional to the problem being solved.
- **Transparency**   The logic by which the regulation addresses the problem must be stated. It must not introduce additional elements that do not relate to the problem. This ensures in principle that regulatory acts cannot be the product of a political bargaining process.
- **Non-discrimination**   Regulation must not discriminate between different players except where there is an objectively justified and revealed reason for it. In particular and in the European context, there must be no difference of treatment in a member state between organisations based in that and in another member state.

3.1.6   Relevant adjacent laws

Much relevant adjacent law has a bearing upon telecommunications regulation and may be referenced in telecommunications legislation and regulatory instruments. Typical relevant legislation includes the following.

- Telecommunications 'code powers' for the obtaining of wayleaves and rights to construct infrastructure.
- Planning and environmental control laws, supplemented as necessary for national parks, public amenity sites, areas of outstanding beauty, historical conservation areas, railway property, industrial sites, energy generating plants and offshore installations.
- Data protection and privacy legislation.
- National security laws regarding the privacy (and interception) of communications.
- Administrative law e.g. the UK's 1978 Interpretation Act.
- Various consumer and trading standards laws.

- Public procurement laws.
- Regulations for the approval of terminal apparatus.

3.2   Styles of regulation

3.2.1   Introduction

Industry regulation may follow a number of differing styles, and a regulator must be aware of the choices being made and why. Regulation may be pursued by means of ex-ante and ex-post rules as explained above. It may have a greater or lesser degree of **intrusiveness** in the day-to-day operation of regulated companies. The exercise of authority by an external regulator is only one of a continuum of options for meeting the need for regulation. These, in increasing order of severity, are listed below.

- **Zero regulation** (let the market decide).
- **Self-regulation**.
- **Co-regulation**.
- **Formal regulation**, or **external regulation**.

3.2.2   Intrusiveness of regulation

Intrusive regulation takes place if the regulator intervenes often and minutely in the day-to-day running of a company's business. Regulation is less intrusive if the regulator lays down more general principles, leaving the company choices over its methods of meeting them. Examples of intrusive regulation would be the dictation of technical solutions, mandating of service quality, control of myriad individual service prices or specification of the amount of air conditioning an operator's exchanges must have. Intrusion is undesirable for three reasons, and the wise regulator will seek to minimise it as far as possible. First, an intrusive regulator builds, inadvertently or otherwise, a role as a proxy manager of the industry, weakening the freedom of operators to run their businesses in the way they think best. It may not be able to bear this amount of responsibility. Second, in intruding, a regulator may increase the dependence of the industry on its existence, the opposite of the aspiration to let competitive market forces rule. Third, the more intrusive the regulation, the greater will be the cost of administering it. Intrusion can be lessened by the use of ex-post rules and by regulating price baskets instead of individual prices. Development of solutions by co-regulation and self-regulation as described below reduces intrusion very effectively. Saying that intrusive regulation is a bad thing does not imply a view that good regulation will have minimal impact on companies in the market. Regulation should be pervasive rather than intrusive, so allowing operators to manage their affairs, as they know best but within the spirit of the regulation.

3.2.3   Self-regulation

Pure self-regulation applies when the players in an industry regulate themselves. For many, this might be their first choice, although smaller players may feel more comfortable when there is an overseeing watchdog. A preference for self-regulation may have suspect origins, when companies want to make pretence of care while in reality looking after themselves, or when big business egos resent other agents no matter how valid their roles. Nonetheless, if it is practicable, self-regulation has benefits. First, companies usually know their businesses better than an external regulator. Second, external regulation has costs that must ultimately be borne by the industry and its consumers. These costs are minimised the more the regulation is internalised. Finally, an external regulator implies hazard, because a regulator is an additional agent player who can get things wrong. Self-regulation is most effective when there is high convergence of interest between stakeholders, and least effective when there are strong commercial conflicts.

3.2.4   Co-regulation

Co-regulation is a hybrid option in an industry that has an external regulator, and forms a halfway house between formal and self-regulation. Instead of formally regulating from outside, the regulator joins the industry players to work with them, as a participant in a joint forum. The regulator may engage with a particular issue to confer authority on an industry initiative, to facilitate collaboration, to represent its position or to seed discussion. This may reduce the need for later intervention. While potentially very successful, co-regulation is vulnerable to role ambiguity and the differing player perceptions of what the others will do and will expect of them.

3.2.5   Formal regulation

An external regulator such as a National Regulatory Authority may deploy any of the four models (formal regulation, co-regulation, self-regulation and zero regulation) in resolution of specific issues, and should choose in the interest of appropriateness and cost-effectiveness. Figure 3.2 below shows diagrammatically how these regulatory models inter-relate, and illustrates a possible route-map for migration of regulation as markets become more competitive.

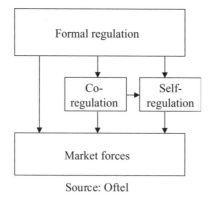

Source: Oftel

Figure 3.2   Possible changes in regulatory style with the development of competition.

3.3   Instruments of regulation

3.3.1   Authorisation

Fundamental to any system of regulation is a system of **authorisation**, where an undertaking needs permission to participate in a regulated market. Embedded in primary legislation, the system of authorisation contains **conditions**. These conditions are the rules, and players in the market must abide by them. Regulation takes place through these conditions. Breach of a condition leaves the company concerned liable to enforcement action, and, in the extreme, to expulsion from the market. Authorisations may be of two types: **specific authorisations** and **general authorisations**. Specific authorisations, frequently known as **licences**, are granted to named organisations. General authorisations permit any company to enter a market subject to its obeying the rules. There should be no necessary administrative act to allow them to start, although the authorisation may contain requirements for registration or notification of the activity.

The European regulation model is currently based on specific licensing. However, limitation of the number of licences by regulatory edict is contrary to the spirit of a liberalised market and should not be practised unless there are sound reasons, for example the radio spectrum limitations that bear on mobile communications markets. Earlier European Directives prohibit limitation of licences other than for objective reasons, and the latest European Directives, which come into effect in July 2003 (see the Appendix), represent a move towards a general authorisation regime. **Licence conditions** provide the framework and embody the rules of regulatory intervention. Asymmetric regulation[3] is implemented by the simple means of having different licence conditions for different players. The licences of the incumbent operators are therefore of utmost importance in the European regulatory model.

The USA's system of regulation is based on general authorisation subject to the **orders** issued by the **Federal Communications Commission** (FCC). Under a general authorisation regime, the regulatory authority must designate certain players as being dominant or having significant specific market power (terminology varies). The rules or orders are conditional upon the market power designation of a company, thus implementing asymmetry.

3.3.2   Licence conditions, rules and orders

The licence conditions, rules and orders by which operators must abide, contain a collection of obligations, prohibitions, conditional permissions and ex-post stipulations, that give effect to regulation. Authorised operators are free to participate in unregulated activities, though subject to conditions that prevent them from exploiting advantages gained as an SMP player in a regulated business to gain an unfair competitive position in an unregulated business. The typical contents of a set of conditions or rules were set out in Section 2.4 under the heading 'The rules of regulation'.

The UK has two licence types, **specific licences** and **class licences**. These correspond to specific and general authorisations. A specific licence is issued to a named company for a determined period of time. Such a licence confers privileges and so has economic value. Specific licences may be of many types depending on the type of player, for example as full facilities operator, cable operator, long distance carrier, value-added service provider, reseller with or without switching, and so on. A class licence refers to a general activity such as the operation of a corporate private network, laying down conditions under which anyone who wishes to pursue the activity can do so without needing to apply for a specific licence. Most regulators issued limited numbers of all types of licence in the early days, partly for caution when newly liberalised markets were in an experimental phase, and also to provide some entry assistance, protecting new players from early additional competition.

Regulation is a living and ongoing process, which must adapt to changing technology and market circumstances. A primary medium of regulatory evolution is via licence or rule amendments. Normally, a regulator cannot impose these unilaterally (except for wholly deregulatory amendments) but must negotiate them with the players after consultation with stakeholders including consumers, government and the other operators. Rule amendments need not necessarily add to regulation, but may subtract from it, for example when regulation reacts to a market that is judged to be becoming more competitive. When the UK regulator, Oftel, introduced an ex-post fair trading condition into BT's licence in 1996, it was able to delete 14 ex-ante conditions. Where agreement cannot be reached, either party may refer the proposed amendments to an arbitrating body. This may take the form of a judicial review in the courts, or a referral to a general competition authority, in the UK's case the **Competition Commission**. A referral might address a point of law, a point of procedure, or the substance of the proposed change.

Referenced documents are references from the rules or licence conditions, and these have the same force as those conditions. A typical example would be the **National Numbering Conventions**. Changes to referenced documents are equivalent to rule amendments and must be pursued as such.

3.3.3   Determinations and consents

Determinations[4] and **Consents** are instruments of ongoing regulation that are provided for in rules. Determinations in effect elaborate rules, and the regulator is permitted and sometimes obliged by the rules to make determinations. In some areas of regulation, price control for example, determinations may be the principal medium by which regulation is enacted. A determination may be generated at the regulator's initiative, be requested by an authorised player, arise from a breach of a rule or licence condition or follow an allegation of anti-competitive behaviour. Typical determinations include the periodic review of price controls, the resolution of quality of service issues, or the determination of the terms of trade between two operators who have failed to reach bilateral agreement. Determinations are usually anticipated in rules by clauses such as, '. . . as the Regulator shall determine,' or, 'unless the Regulator shall otherwise so determine'. A consent is a negative determination by which an authorised operator is formally relieved from the need to comply with a requirement, and consents are typically anticipated by clauses such as, 'unless the Regulator shall otherwise so consent'. Consents and determinations have equal power with the rules or licence conditions themselves, so breach of a determination or consent is equivalent to a breach of the governing rule. Like rule amendments, determinations and consents may be challenged and referred for review in the courts or by a general competition authority. Because they have legal force, determinations and consents are, like rules themselves, drafted in legal language.

3.3.4 Guidelines

Guidelines are informal documents that do not in themselves possess a legal status and are usually presented in user-friendly language. However, they are important in that they convey useful information. They have authority in the sense that they reveal the thought processes and criteria that may be brought to bear in deciding whether a breach of the rules has occurred. Guidelines might cover such things as standards for presenting tariff information, an understanding of what is meant by 'reasonable interconnection' or technical interface standards. Self-regulation or co-regulation initiatives typically result in the issue of guidelines.

3.3.5 Consultation

Regulators must consult widely in arriving at strategic policy, regulatory procedure, rule amendments and other instruments such as determinations and guidelines. The stages in a consultation typically include:

- notification and invitations to contribute;
- publication of **consultation papers**;
- conduct of consultative meetings and consultations in writing, with iteration if necessary;
- issue of a memorandum to accompany the final legal instruments.

Consultations give the industry both ownership and pre-knowledge of decisions. They enrich regulatory decision-making by involving the wide knowledge and experience of all those consulted. Invitations to contribute may be issued as appropriate to stakeholder interests such as government departments, authorised operators, consumer groups, regional representative bodies or more generally to any member of the public who wishes to contribute.

A requirement to consult various bodies may be enshrined in law and in rules. Consultation may appear at first sight to be nugatory, but this is not so. A climate of transparency and openness makes it difficult for a regulator to ignore genuine inputs, notwithstanding that consulted parties acquire no formal powers in the making of a decision.

3.3.6 Judicial review

The actions of a regulator are subject to appeal, either by a body specifically appointed for the purpose, or by judicial review in the courts. An appeal may be on the grounds of law (illegality), on the grounds of procedure (misadministration), on the grounds of logic (irrationality) or it may challenge the substance of the action.

3.4   Enforcement

Regulators have a variety of powers at their disposal for dealing with rule breaches, as shown in the following list. Concurrently they have policing powers to search for, identify and investigate breaches and anti-competitive behaviours, whether on their own initiative or in response to allegations and complaints.

- Licence revocation.
- Fines.
- Compliance orders.
- Exposure to litigation for damages.
- Publication, 'naming and shaming'.

Regulators' powers are strong but not as absolute as some people may imagine. The consent of the governed is as important a facet of telecommunications regulation as in many other areas of law. Regulators' powers are bounded by legislation. Where regulation is distributed between a national regulatory body and other bodies, the exercise of enforcement powers may require agreement with other bodies that cannot be taken for granted. The limitation of regulators' powers by government is a natural and cautious approach to liberalisation, as it provides some checks and balances before the government has seen how the regulator will, in practice, develop its role and assert its powers. However, even had a regulator the powers of an absolute monarch (that is, a wide remit, no accountability, no liability to appeal and draconian enforcement powers), it would still need the co-operation of the industry to make many things happen.

Licence revocation is an extremely strong and disruptive power. It would be ineffective were it the only power, since, being disproportionate to most breaches, it would not in practice be used. **Fines** are probably more effective as a threat than by their actually being imposed. Large fines would seriously affect the costs borne by the industry and ultimately consumers. If fines were minimal, operators could in principle adopt a culture that viewed them as operating expenses, while the regulator's aim is compliance, not punishment. It is interesting to note that some regulators do not possess the power to fine, as was the case with Oftel in the UK until 1999, although other countries' regulators have made use of fining powers.

A **compliance order** is a formal notice served on an authorised operator to comply with the rules. Many countries have a hierarchy of orders, starting with **provisional orders** leading later to **final orders**. Orders are, like other instruments, subject to a right of appeal. Regulators may have pre-emptive powers to require immediate compliance where issues of public safety and basic service integrity are at stake. Often, compliance orders achieve effect by the simple threat of inconvenience of ongoing proceedings. Lawsuits and regulatory action can consume a great deal of an operator's management time, and may cause them to take their eyes off the important job of running the business. They may also prejudice the operator in the eyes of financial markets, potential alliance partners and foreign regulators, who may be judging its eligibility to make acquisitions in their own countries.

The exposure of an operator to civil litigation for damages by an injured party, perhaps by another operator that alleges it has suffered from unfair competition, is a powerful deterrent to breaches of rules and licence conditions. This is in practice the primary medium of enforcement in some countries, especially where there is less or no sector-specific regulation of telecommunications. Informal powers such as publication of offences, comparative prices or league tables of service quality may be effective in certain types of case. The procedural burdens of non-compliance serve in practice to keep operators in line with the rules and their licences simply by making life very difficult if they depart from them.

3.5 Funding and accountability

3.5.1 Funding

External regulation is an expensive activity whose considerable costs must ultimately be borne by the industry and its consumers. The experience of the United States after the 1996 Telecommunications Act showed that the inflation-adjusted annual budget of the Federal Communication Commission (FCC) in 1996–2001 was higher by 37 per cent than over 1981–1995, while the number of output pages nearly tripled[5]. The direct cost of the national regulatory authority itself, typically millions, tens or even a hundred millions of pounds per annum, is only the first item in a list that includes the regulatory affairs people that authorised operators must have, the costs of compliance, and the costs of being able to demonstrate compliance. Appropriate and proportionate regulation is, therefore, very important, while the development of self-regulation and co-regulation are preferable alternatives to a burgeoning regulatory authority.

Governments assume primary responsibility for the funding of their NRA but usually pass this onto the industry in the form of fees. The setting of fees is no trivial matter, as there is no simple formula based on turnover, profit, capital employed or investment that would provide fair shares for new and established players at all stages of their business cycles. Because the amounts involved are usually tiny in relation to the size of the players, informal determinations are often achieved without too much difficulty. A criterion used in the Netherlands and being considered for the UK is to raise a fee proportional to allocated numbering capacity. This is a fairly accurate measure of an operator's size, and is unlikely to distort management strategies through attempts to minimise it. It is important, and required by European Directives, that fees collected reflect only the cost of regulation and are not used for other purposes.

3.5.2 Independence and accountability

National regulatory authorities must have independence from the companies being regulated. While a government-established regulator possesses obvious independence from private companies, there are conflicts of interest when players are owned, part owned or in any other way sponsored by government. The German government, for example, had a problem of independence when the Post and Telecommunications ministry[6] responsible for the incumbent Deutsche Telekom also undertook regulation. This was resolved in 1998 with the setting up of the independent regulator Regulierungsbehörde für Telekommunikation und Post (RegTP).

Government-appointed regulatory bodies must be accountable for their actions. Typically, they answer either to a government ministry other than the telecommunications ministry, or to the government or parliament in general. The Netherlands regulator, OPTA[7], and the German regulator RegTP answer to government ministries[8]. The British regulator is answerable to the national parliaments, and may face questions from any part of government or from parliamentary committees. The United States' federal regulator, the Federal Communications Commission (FCC), is independent of government but must implement Congress's policies. Accountability to government or parliament is no mere token accountability, since parliamentary committees and commissions of enquiry are capable of posing very searching questions that require public answers.

3.6   National frameworks

3.6.1   UK and Europe

The UK and European model of regulation is based upon sector-specific regulatory bodies and the individual licensing of industry players. It has been widely adopted outside this region, and is typical of many markets that progressed from state ownership to liberalisation in the 1980s and 1990s. Many of these countries had little specific telecommunications law before liberalisation. Laws that existed were typically general or antiquated laws, placing the control of telecommunications in the hands of the government. Germany had an article in its constitution to this effect. National laws establishing sector-specific telecommunications regulators have been enacted alongside liberalisation in many countries, for example the UK (1984), France (1990 and 1996), Australia (1991), Canada (1993), Sweden (1993) and the Netherlands (1994).

The UK 1984 Act appointed a regulator, the Director General of Telecommunications, giving him or her jointly with the secretary of state (i.e. the government department, in this case of Trade and Industry) a duty [4]:

(a) to secure that there are provided throughout the United Kingdom, save insofar as the provision thereof is impracticable or not reasonably practicable, such telecommunication services as satisfy all reasonable demands for them including, in particular emergency services, public call box services, directory information services, maritime services and services in rural areas; and
(b) without prejudice to the generality of paragraph (a) above, to secure that any person by whom any such services fall to be provided is able to finance the provision of those services.

Section 3(2) goes on to require the regulator to exercise his or her functions in a manner that is best calculated *inter alia* to[9]:

- promote the interests of consumers (including the old and disabled);
- maintain and promote effective competition between persons engaged in commercial telecommunications services.

Market entrants who wish to provide public telecommunications services require an authorising licence. This contains the rules through which regulation takes place. The specific licensing of individual players permits asymmetric regulation. Those deemed to have dominance or market power have different licence conditions and thereby receive more stringent regulation than the generality of industry players.

The basis for the enforcement of regulatory instruments in the UK is as follows. Breach of a licence condition is in itself not a criminal offence. However, the regulator may issue a compliance order, and in case of non-compliance, three things may follow. Failure to comply would be a breach of duty by the licensee, leaving it liable to any person who suffered loss or damage in consequence of the breach. If the order were backed with a court injunction, failure to comply would place the licensee in contempt of court, which may be punished by fine or imprisonment. In cases of persistent non-compliance, the regulator may revoke a licence. After this, the former licensee would be guilty of operation of an unauthorised system, which is a criminal offence.

Judicial review of regulatory decisions was (and is) available according to normal UK law and practice for the challenge of the actions of government. This permits a review on the grounds of law (illegality), on the grounds of procedure (misadministration) or on the grounds of logic (irrationality). This did not originally allow for review of the *substance* of a decision, though this has since become possible in conformity with European directives.

3.6.2   USA

United States regulation differs greatly in its form and history from the UK and Europe. US telecommunications, historically provided by private companies, were regulated by the 1934 Telecommunications Act, which had a public interest remit. This Act created the **Federal Communications Commission** (FCC) to take regulatory oversight for inter-state and international communication. For much of the time since the Act, the FCC permitted the monopoly of AT&T as a 'common carrier', operating under price control and with universal service obligations[10]. The FCC is required to implement government (Congress) mandates, which in practice have varied from time to time between broad policy and very detailed management (for example of premium-rate information services). The FCC does not have a licensing system as in Europe, but regulates by the promulgation of general industry rules known as 'orders'. Regulation may evolve by legal challenge in the courts, and the development of regulation in the USA shows greater use of litigation than in Europe.

The act of deregulation of US telecommunications, paralleling the UK's Telecommunications Act, was the **Modified Final Judgement** (MFJ) of 1982, taking effect in 1984. This was the culmination of litigation that began in 1963 with a petition by Microwave Communications Incorporated (MCI) to be allowed to provide inter-state communications in competition with AT&T. The MFJ removed the AT&T monopoly and created seven local operators, the **Regional Bell Operating Companies** (RBOCs). The RBOCs continued to enjoy local monopoly within their areas, but had to send inter-city calls via the competing long-distance carriers, to whom they had to provide equal access[11].

A **Public Utility Commission** (PUC) in each state performs regulation at the state level. Many regulate other utilities besides telecommunications, and most employ some form of price control. Their commitment to local competition has shown considerable variation, with New York and Illinois among the most liberal in the early days. The legal doctrine of federal pre-emption prevents states from legislating inconsistently with federal (FCC) policy on matters that are entirely within the field of Congress, but nonetheless there is scope for tension between federal and state regulation. Companies whose operations cross state boundaries, in practice most of them, find it necessary to negotiate with diverse regulators.

With the 1996 Telecommunications Act, Congress replaced the policy framework of the 1934 Act, and decreed a more competitive framework than either that of the 1934 Act or the 1984 MFJ. It:

- overturned the RBOC monopoly for local services;
- declared invalid any state regulation having the effect of restricting the entry of competitors into interstate telecommunications markets;
- permitted the RBOCs to provide interstate service, though subject to satisfying a checklist of things an RBOC must do in encouragement of local competition (for example, to provide number portability);
- set rules for incumbent local carriers and new market entrants.

3.6.3   New Zealand

New Zealand has followed a very different regulatory path from most countries by proceeding without a national regulatory authority. Market entry is open under a general authorisation, and operators are governed under general competition law, in this case the 1986 Commerce Act. New Zealand's experience has shown that this method of regulation makes high use of litigated settlements. As an example, an interconnection dispute between Clear Communications and Telecom New Zealand (the former monopolist) progressed to New Zealand's highest court, the Judicial Committee of the Privy Council, which is based in London. A national regulator can act as a useful proactive agent when initiatives need to be pursued, and New Zealand has possibly felt the lack of this. A recent government inquiry has recommended change, to light touch regulation by an industry-specific regulatory body [5].

3.7   Dangers of regulation and operational aspects

3.7.1   Regulatory hazard

Regulation carries fundamental powers over the regulated industry. This power is hazardous, as it can distort markets as well as benefit them. Regulatory power, if exercised inappropriately or unwisely, may wreak great damage on the industry and the user community it serves. It follows that regulatory actions must be subject to impact appraisal, and take due consideration of the likely or potential consequences that may follow from rulings. Perhaps the best-known recent example of regulatory failure was in the electricity supply industry of California in the late 1990s. Retail price regulation, presumably enacted to serve the consumer interest, did not relate to a deregulation of wholesale input price levels, thus permitting wildly unbalanced prices to develop. Adverse factor price movements depressed profits, so forestalling investment in the essential utility and eventually leading to widespread power outages. It did not need a sophisticated or expert analysis to foresee this.

The best defence against inappropriate regulation lies in the underlying philosophy of administration. As explained in Section 3.1.5, this should ideally rest on stated goals, proportionate actions that address only the goals, transparency and recourse to courts of appeal backed by accumulated jurisprudence. These are the principles underlying competition law. Distortions are likely in the following circumstances.

- The regulatory authority does not follow a market-analytic approach but adopts *ad hoc* though well-intentioned measures, for example, 'We will have three competitors for our incumbent' or 'Anything that can be unbundled must be'.
- The regulatory authority is captured by a political desire to produce certain outcomes. Governments may wish to retain control despite theoretical commitment to free markets.
- The regulator neglects longer-term market development (for example, investment or growth of sustainable competition) in favour of popular short-term outcomes such as price reduction.
- Regulatory decisions are subject to industry lobbying and political bargaining processes.

3.7.2 Regulatory paralysis

Regulation carries the risk of paralysing the industry should the strategic thrust of the regulatory agency become uncertain and remain so for any significant period of time. Not knowing who will be allowed to play in a new business or the terms of trade therein will deter investment decisions. Sometimes, industry players will find that they can serve their own interests by paralysing regulation. The division of the regulatory function may, if poorly designed, lead to disagreements, power play and slow decision making. Opportunistic companies may find they can play regulatory jurisdictions against one another.

3.7.3 Skill requirements

A national regulatory authority needs to build up an impressive set of expertises, some of them at a senior level and which may be rare in the marketplace. These include the following, all of them with special emphasis on their application to the telecommunications sector:

- Lawyers.
- Economists.
- Engineers.

- Administrators.
- Accountants.
- Statisticians.
- Market researchers.

As in any organisation, a mix of temperaments is required. These range from the rebels and visionaries who are willing to think the unthinkable, question the obvious and foresee the future, through the analysts who can penetrate issues quickly and assimilate different viewpoints, to the technical specialists who can work reliably and accurately. The situation was complicated, at least in the early days, by the inevitable learning curve when people individually and collectively addressed the new activity. They simultaneously faced urgent demands from government and the industry to deliver clarity, action and results. There is a potential problem when a regulatory authority employs career civil servants who pass through it as a posting stage on their journey of acquiring varied experience. If, say, three years were the average stay of such a person, he or she would often have reached productive efficiency at the very point of departure, taking hard-to-replace knowledge with them. A pool of high calibre managers released by downsizing at national monopoly operators has helped, although it is important not to obtain too many people from this source lest there be an inherited cultural tie.

3.7.4   Information asymmetry: the knowledge gap

All regulators face an enduring problem, which is that the companies they are trying to regulate know much more about the business than they do. A major advantage, however, when facing the information asymmetry problem, is that the regulator has and must exercise discretion in applying rules and conditions, and in deciding what regulation is appropriate. This gives it the power of the initiative. If its function were purely analytical and bureaucratic, then it would not be able effectively to counter the superior knowledge available to the dominant player.

Obviously regulators must develop certain key expertises in-house, but they will also depend on longer and shorter-term access to external sources of help. Suitable sources of expertise include contract staff, consultants, academics, professional networks and the industry itself. This last is an important source when industry and regulator can find a common objective in pursuing particular topics. Co-regulatory initiatives help, therefore, to solve the knowledge gap. Lawyers, with their ability to master a 'brief', are a valuable resource.

Contract people are a solution to the problem of obtaining key expertises that may only be needed for a bounded duration, such as for a national number change or a strategic investigation. These may be needed at a time when cyclical demand has made them scarce, so a regulator may have to pay higher rates than civil service norms. Finding the right people can be difficult: on the one hand someone might bring excellent intellectual penetrating power but need a little learning curve in the topic on hand, while another with the exact track record may bring a disappointing pre-commitment to the way something was done in his or her last project.

Consultants ought in theory to be an ideal resource, since this sector has blossomed in size with the more flexible labour markets now seen in many countries. Consulting is a sensitive activity requiring rapport and trust because the client pays for knowledge, but cannot know in advance what he or she does not know. There are mature consultants around with the desire, ability and yet humility to be able quickly to identify with stated and unsaid client requirements, and to contribute faithfully, accurately and quickly. Nonetheless, finding people of such worth may be more problematic. Those possessing the finest qualities may only be identifiable after employment. Some will come from cultures that treat an assignment as a failure unless it opens the door to further consultancy.

3.7.5   Regulatory capture

The phenomenon of regulatory capture takes place when the regulator identifies itself too closely with, and so serves the interests of, the industry players. It would be especially relevant in telecommunications should the regulator align itself consciously or otherwise with concerns of the former monopolist. Cases abound especially in the USA in railways, road transport and the airline industries, where regulation has protected the established companies from competition in return for price control and supervisory functions. Political capture, resulting perhaps in commitment to certain market outcomes, is a possible disturbance to a regulatory process.

Capture by the regulated industry, and in particular by the monopolist, can arise through a number of mechanisms. It was partly with such factors in mind that Australia in 1997 disbanded its NRA (Austel) and handed regulation over to the general competition authority, the Australian Competition and Consumer Commission (ACCC). The general body with the wider remit was thought to be much less vulnerable to capture by any one of the many industries with which it dealt. A sector-specific body, the Australian Communications Authority (ACA), also exists with certain technical and consumer protection responsibilities.

3.7.5.1   The information gap

Regulation is in danger of capture should it depend too closely on the monopolist for the input of knowledge and data about the industry. The use, therefore, of external sources of knowledge and the development of co-regulation involving the monopolist's peers is essential. The existence of exceedingly complex regulations, as was and is the case in the UK with certain BT licence conditions, sets up vulnerability, since the regulated player may well possess vastly superior skills in interpreting and operating the conditions.

3.7.5.2   Personal allegiances and career prospects

The risks of regulatory capture are magnified if the people at the regulatory agency have strong personal allegiances with those working for the monopolist, and if they depend for ongoing career prospects on being well regarded in the industry. The first risk can be controlled by developing a strong *esprit de corps* at the regulatory agency, and by recruiting people from a mix of backgrounds. The latter depends on there being adequate career openings for regulatory people apart from within the industry or with its monopolist.

3.7.5.3   Intellectual factors: 'self-inflicted' capture

Regulators normally pride themselves on their logic, fairness and transparency in executing regulation. This can, however, lead to weakness should fear of 'losing the argument' engender timidity of action. The regulator may then fail to use even powers that are available, transfixed while the regulated player takes comfort in prolonging debate. Regulators will naturally spend much time working with people from the monopolist, with whom they will develop professional relationships. Should the regulator and regulated player reach accommodation over a contentious and difficult issue, then a subtle capture effect may follow over time as the regulator abides by an agreement in which it may have invested some pride of achievement.

3.7.5.4   Political lobbying

The lobbying of a regulatory agency to 'look after' certain vested interests or to secure favoured 'outcomes' will always be a problem for any regulatory regime. Transparent administration and separate accountability of regulation are necessary antidotes here. The division of regulatory functions between different bodies, such as the UK's separation of the licensing function (at the Department of Trade and Industry, DTI) from the supervision and enforcement function (at Oftel, the NRA), may provide a check against capture, or create a vulnerability to it.

3.7.6   Regulatory culture

Like any organisation, a regulatory agency will develop a culture. This will arise from both the job it has to do and the qualities of the people who lead it. Whereas one regulator may value an ego-less approach to its operation, another may position its leader as a figurehead or 'industry champion'. A typical regulatory authority probably likes to believe it has an analysis-centric culture, where it applies stated criteria to defined problems and so produces rules that can be proven to be correct, beneficial and appropriate. This view treats its role as a classical bureaucratic function, such that one might expect equally competent functionaries to arrive at the same outputs regardless of personality or preference. This is not a wholly appropriate model of regulation, however, there being a need for constructive discretion. Experience shows that telecommunications regulators will need to cultivate a variety of working styles to face specific situations. The study in Reference 6 shows that besides the bureaucratic approach with linear progression towards an endpoint, regulators may exhibit two other operating styles. The bargaining-diplomatic style applies when a substantial learning curve implies a discovery process within the problem-solving path. There will then be plenty of iteration between regulator and stakeholders. The chaotic style comes into its own when the problem space is so uncharted that not only is there discovery within the process, but also *ad hoc* discovery of the process. Thinking may then be lateral, and the problem solving path unpredictable.

3.8 Notes

1 This list excludes laws that bear on all and sundry, for example those concerning companies, taxation, finance and accounting, health and safety, planning and the environment, trading, equal opportunities, human rights and employment.
2 Author's paraphrase.
3 Asymmetric regulation, introduced in Section 2.2.2, is the application of different rules to different companies. It is normally employed to permit tougher regulation of dominant companies.
4 This terminology follows UK practice. Other countries may use other terms, although the concepts presented are generic.
5 This data is taken from the FCC Record and quoted in Reference 7.
6 Bundesministerium für Post und Telekommunikation.
7 Onafhankelijke Post en Telecommunicatie Autoriteit.
8 The ministries are the Ministry of Economic Affairs (Ministrie van Ekonomische Zaken) in the Netherlands with a constitutional independence for OPTA, and the Federal Ministry of Economics and Technology (Bundesministerium für Wirtschaft und Sitz) in Germany.
9 Author's paraphrase.
10 These words do not appear as such in the 1934 Act, though the phrase 'to make available, as far as possible, to all the people of the United States' is usually interpreted as a basis for universal service obligations.
11 Equal access is a facility for the owner of any access line to nominate a choice of inter-exchange carrier. Calls are passed to the chosen IXC without any explicit action such as the dialling of an access code.

3.9 References

1 WALDEN, I. and ANGEL, J. (Eds): 'Telecommunications Law' (Blackstone Press, London, 2001)
2 LONG, C. D. (Ed): 'Global Telecommunications Law and Practice' (Sweet & Maxwell, London, 2000)
3 'The benefits of self and co-regulation to consumers and industry'. Oftel, July 2001
4 UK Telecommunications Act 1984, Section 3 (1) (HMSO, London)
5 'Ministerial Inquiry into Telecommunications – Final Report' (New Zealand Government, Wellington, September 2000)
6 HALL, C., SCOTT, C., and HOOD, C.: 'Telecommunications regulation – culture, chaos and interdependence inside the regulatory process' (Routledge Advances in Business and Management Studies, London, 2001)
7 SIDAK, J. G.: 'The failure of good intentions: the collapse of American telecoms after six years of deregulation', Beesley Lecture on Telecommunications, 1st October 2002 (Institute of Economic Affairs and the London Business School, London, 2002)

Chapter 4

Regulatory strategy and price controls

4.1 Appropriate regulation and the need for discretion

Regulation is an activity requiring a great deal of discretion. It is not simply a linear, analytical process, nor is it a purely bureaucratic or policing role. An industry regulator is a constructive agent that must regulate appropriately, since the potential of regulation to bring consumer benefit carries the counterbalancing hazard of inflicting damage on an industry. A regulator within a market economy is not a market manager, and should not attempt to be one. Regulatory market management is in the long run impossible, since no regulator or government has the power to compel individuals or companies to enter a market or make investments. What the regulator should do, however, is to ensure that the market has the maximum potential to operate freely. In so doing, it must address market failure and intervene to curtail the abuse of market power. A regulator's guiding principles, a list of both positive and negative propositions, might possibly read as follows:

- to create a thriving, viable industry, delivering quality, value-for-money services;
- to resolve market failure and allow competitive market forces to operate as efficiently as possible;
- not to attempt to engineer specific outcomes;
- not to engender competition for competition's sake.

Regulatory actions need to be subjected to regulatory option appraisal. This may involve methods such as cost-benefit analysis, regulatory impact analysis and consumer impact analysis. These may take place explicitly, or implicitly through consultations with stakeholders. All regulatory actions have costs, at minimum the costs of regulation and compliance. They often impose costs on the industry and ultimately the consumer, as is the case, for example, in the requirements to interconnect and to provide number portability. Balanced against these costs are the consumer benefits. Wealth-creation may result from greater competition. There may be a reduction of waste, when an inefficiency caused by market failure is removed. In all cases, however, the benefits must exceed the costs, and regulators will tolerate a market failure whose risks or costs of removal exceed the harm it causes.

A regulator must consider the risk entailed in a proposed intervention, including and in particular the effect it might have on long-term investment. It is possible to encourage inappropriate competition. The stimulation of new market entry may not be beneficial if it lifts risk and uncertainty to the point of raising the industry's cost of capital or depressing long-term investment. Excess competition and tight price control may generate popular price reductions in the short term, but consumers would suffer ultimately if these had the effect of discouraging investment. Forcing the incumbent or other players to make investments that might be stranded (that is, there is a risk of premature obsolescence) might ultimately raise industry costs to the detriment of the consumer.

In nurturing competitive markets, regulators should not normally engage in proactive and affirmative entry assistance. They must, however, tackle entry barriers. Competition-friendly regulation is valuable only to the extent that it serves consumer well being. A competitive market is of value to consumers only if the suppliers in that market are profitable undertakings. If a market can sustain viable competitors, then they will come forward by themselves, as long as the regulator ensures that companies with the market power to prevent them unfairly do not do so. In many markets, there is an optimum number of players, a number that a free and open market will find for itself. If regulation resulted in the entry of companies whose existence depended solely on the operation of regulation, then that regulation would not be appropriate.

4.2   Market strategies

4.2.1   The need for a regulatory strategy

A regulator must develop a strategic understanding of the market it is regulating. This is not in order to practice market management, but because an understanding of the factors driving a market allows the regulator to appreciate the market structures that might develop. Against this understanding, it is possible to test the likely results of intervention and so decide the most appropriate regulatory intervention, if any.

Technological change, which is rapid in the information and communication industries, is a major driver of regulatory policy. This is because technology changes the economics of an industry by:

- introducing new services;
- changing the underlying costs of a product or service;
- shifting the cost profile of an industry, for example from being capital intensive to being current-cost intensive or vice versa;
- affecting the scale economies within an industry.

4.2.2   Is telecommunications a natural monopoly?

Central to any regulatory strategy for the introduction of competitive supply to a monopoly industry is an understanding of whether the monopoly in question is a natural monopoly or not. A natural monopoly exists, as explained in Chapter 1, whenever there are unused scale economies at production levels such that a single supplier could supply all or the bulk of the demand in the market, or where it would be obviously undesirable to duplicate a large supply infrastructure. The telecommunications services industry is a composite one, so it is quite possible for one part to be a natural monopoly while another is not.

Long distance carriage of voice and data in telecommunications is not a natural monopoly, as has been amply demonstrated by the flourishing of competitive supply in national and international markets. It was, however, a natural monopoly a hundred years ago, and this led in the early 20th century to monopolistic consolidation of the pioneering suppliers of the late 19th century. Technological development has since caused a revolution in the economics of voice and data transport, and its cost continues to fall. A circuit from, say, London to Glasgow using un-amplified copper pair technology in around 1900 would have needed two wires as thick as a pencil, using 300T of copper costing about £315,000 at today's prices[1]. In contrast, an optical fibre thinner than a human hair can carry a phone call for every man, woman and child in a city the size of Leeds, or 300 uncompressed TV channels[2]. As well as this dramatic reduction in the input factor cost, there is a scale economy whose effect was severe with the small call volumes of 1900 but is largely insignificant with the large traffic flows of the 21st century.

This scale effect, known informally as the **Erlang effect**, is named after the engineer A. K. Erlang (1878–1929) who was influential in the development of the statistical theory of telephone traffic. The economy of scale arises from the increased efficiency with which telephone calls can be carried when there are large numbers of them between any two places. To provide an acceptable quality of service, a telephone network must ensure that there are sufficient circuits or links on each route so that one will be available whenever someone wishes to make a call. The **grade of service** may be defined as the probability that a call will fail due to a circuit not being available. The lower the number, the better, and 0.002, or one in 500, is a standard planning target for many modern systems. If, for example, there was an average of three calls in progress between two towns, the theory shows that ten circuits would be necessary to ensure a grade of service of 0.002, and as a result, the mean utilisation of those ten trunks would be about 30 per cent. With greater numbers of calls, there is greater chance that one call will be finishing at or near the time of a new call arrival. This diminishes the necessary overprovision and raises the efficiency with which the larger volume can be carried. The Erlang effect is illustrated graphically in Figure 4.1. This plots the efficiency with which circuits are loaded (vertical axis) against the mean number of calls on the route (horizontal axis). For example, for an average of 50 calls in progress between two towns, the graph shows a carriage efficiency of nearly 63 per cent. This implies that 80 circuits would be needed on the route to support the standard grade of service. Naturally, a route must be dimensioned to the needs of the busiest hour of the day rather than a 24-hour average, which further and significantly diminishes the proportion of time a circuit is earning money.

A local access network provides a copper pair to millions of homes and businesses. This network represents a large sunk investment, which has been quantified at around

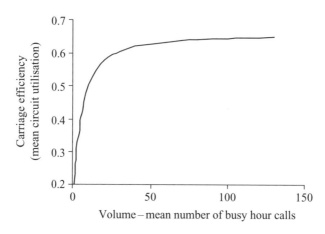

Figure 4.1   Efficiency of carriage of various call volumes.

£800 per line at current prices, giving, for example, BT's access network a replacement value in excess of £20 billion. Local access networks usually fail, in isolation, to earn their cost of capital. It was clear in the UK that no competitor was likely to replicate BT's access network excepting in business-intensive downtown locations. It might appear, then, that local access is a natural monopoly.

This appears to have been the view in the United States at the 1984 deregulation. With the divestiture of AT&T, a market structure was created with seven regional monopolies providing access and local service, the **Regional Bell Operating Companies** (RBOCs), and a liberalised long distance (**Inter-exchange Carrier**, IXC) market. Competition flourished when MCI, US Sprint and others entered the IXC market to compete with the incumbent AT&T. Liberalisation strategies in other countries have taken different positions, however.

In the UK, it was believed that the local network monopoly was contestable by new technology, most notably by cable TV networks. It was assumed that the cable companies could make profit, given the scope economy of combining broadcast entertainment with telecommunications. They were expected to be able to provide telephone access at an incremental cost of about £280 ($433) per line [1]. Wireless loop technologies offered a further avenue for contesting copper loop monopolies, since their cost of provision was expected to fall rapidly in the 1990s from £645 ($1,000) per line to a competitive £320 ($500) [2]. Sadly, the potential of the wireless loop has failed to materialise as hoped, principally because of the difficulty of achieving the necessary subscriber density quickly enough to finance the heavy fixed charge of the base stations. Infrastructure duplication does, of course, inflict an economic penalty on the industry. Countries such as the UK have made, in effect, an act of judgement and of faith, that the dynamic gains resulting from competition would offset this and yield net benefit. The mobile networks also provide a radio access infrastructure, and these also challenge the access monopoly of the incumbent copper loop.

In industries where a clear natural monopoly is acknowledged, such as railways, electricity and gas supply, competition has often been introduced by the division of the former monopolist into a regulated natural monopoly infrastructure company, and one or more types of retail service company. Competing with one another, these buy the basic infrastructure service at a regulated price as an intermediate input to the retail product.

4.2.3   Competitive arenas in telecommunications

Competitors may contest a monopoly market by providing a full range of services, thus imitating the monopolist. This is, however, by no means the only possible medium of competition. Various types of competition for an incumbent telecommunications operator might be as follows.

- National full facilities operators, who have access and long distance networks and so imitate the monopolist.
- Territorial full facilities operators, challenging the monopolist but over a restricted area.
- Long distance carriers, offering long distance service but relying on the access network of the monopolist for bringing calls to the network, and again for delivering them to the end destination.
- Resale operators, who buy monopoly services such as long distance leased circuits and then resell that capacity for individual calls or other managed services.
- Value added service providers, who offer peripheral network services such as dynamic routeing, number translation, Freephone, voicemail and premium rate services.
- Specialist business suppliers, who compete with individual parts of the monopolist's business, such as apparatus supply or systems management and consultancy.
- Information service providers (ISPs), who offer services such as web hosting, electronic mail, Internet access and information services.
- Virtual network service operators, who basically rebrand the service of the monopolist (or another operator), handling customer care and billing, and possibly adding other value. The added value may include 'affinity services', where the service is bundled with something else such as gas supply or financial services.
- Services over unbundled infrastructure, where a new entrant makes wholesale purchase of a right to use plant such as the customer's local loop, and provides competing services over it.

4.3   Market structures

Licensing systems (where they exist) and regulatory rules give regulators considerable power to influence the structure of the industry. This gives expression to the regulator's view on whether or not the market is a natural monopoly, and how it can best be structured to allow competition to develop and flourish. United States regulation at the 1984 divestiture of the AT&T monopoly, for example, took the view as described above that the 'last mile' customer access networks of local telephony operators were natural monopolies. Many countries will have given careful consideration to the vertical separation of their monopolies into service and infrastructure companies. British regulation took the view that the incumbent BT should continue intact in its former monopoly shape (being of course regulated as an SMP player), expecting that in due course full facility operators with local access as well as long distance networks could and would enter the market. The decision by the UK and most other European countries not to divest their monopolies as in the USA followed not only from economic strategy, but also for pragmatic reasons. They had fears whether a break-up of their vertically integrated telecommunications companies would work technically and operationally. It is worth remembering the scale factors that make a single US RBOC larger than most European national operators. Also, they wanted to get strong prices for their monopolies at privatisation, and leave them commercially powerful enough to have a chance of impact as global players.

In the early days, the UK used the licensing system to engineer a duopoly from 1983 to 1991 between the UK's second network operator, Mercury, and the incumbent BT. This gave protection to Mercury against further competition, and no doubt echoed a cautious approach to market liberalisation on the part of the licensing authority (the Department of Trade of Industry). Mercury's privileged position was removed at the Duopoly Review of 1991. This introduced the now current regime, which does not limit entrants to the UK fixed telecommunications market. Cable TV operators, who were permitted to offer voice telephony services via BT or Mercury from 1987, were encouraged from 1993 to offer these in their own right. To encourage their invest-ment in a national access infrastructure, the cable TV industry was regulated under a monopoly franchising system, giving protection to their economy of scope (of TV with telephony). Operators had exclusivity in providing cable TV services in the areas where they held a franchise. Telecommunications operators (all of them, not just BT) were prohibited in their capacity as licensed telecommunications operators from using their lines to compete for TV services delivery for ten years. They could, however, bid for franchises if they wished, and indeed BT obtained some area franchises (which it eventually sold). The duopoly and the cable exclusivity are examples of affirma-tive, proactive entry-assistance. This is fraught with dangers of market distortion and should be undertaken only when considered unavoidable, as in these cases to de-risk large infrastructure investments. Like monopoly, protection can shelter inefficiency.

The UK's cellular mobile telecommunications market opened with a duopoly between Cellnet (at the time part owned by BT)[3] and Vodafone, who received their licences in 1983 and began commercial operations in 1985. Both were deemed to have significant market power[4]. The UK regulator (the Department of Trade and Industry, not Oftel in this case) imposed an industry structure that prohibited either network operator from retailing mobile communications services directly to end-customers. Instead, they were required to supply network services via retail **Service Providers** who handled sales, service, billing and the customer relationship. This industry arrangement has rarely been replicated elsewhere, giving the term 'service provider' a unique meaning in the UK. When a third licensed player, One2One, entered the market in 1993 and a fourth, Orange, in 1994, these were deemed not to have SMP and were not prevented from developing their own retail mobile services businesses.

Market structure concerns did not disappear with the passing of the early days, but remain vitally relevant. A recent example, described in detail in Chapter 9, was the question of whether it would be appropriate regulation to require mobile operators to supply wholesale air capacity to mobile virtual network operators. Regulators need to be sensitive to the risks of over-stimulating competition to an unsustainable level, as this will not result in an efficient market. This can be brought about by unrealistically tight wholesale price control, allowing inefficient players to survive by obtaining use of the incumbent network below its true cost and at the same time discouraging investment in the incumbent's network.

4.4   Price control

4.4.1   Principles of price control

Price control is the classic regulatory remedy for the market failure of monopoly, and provides a force on monopoly prices as a proxy for the downward price pressure that a competitive marketplace would have exerted. It is important to distinguish between the use of regulatory price controls for correction of market failure, and their possible use for other purposes, such as social objectives, consumer objectives, market management or producing outcomes. Any use for these other purposes beyond a marginal scale will be market distorting and so produce non-optimality. Should a government wish to pursue them, it is probably best for the government to enter the market as a customer or joint customer and so make any transfers or subsidies explicit.

Price control may be applied to end-customers' retail prices, to wholesale prices for other operators and service providers, and to interconnect rates, a special case of wholesale prices. Price control is normally subject to periodic review. There are three basic types of price control:

- rate-of-return control;
- cost-based price control;
- direct price control.

Hybrid price controls combining these elements are possible, as for example when a wholesale price is controlled on the basis of avoided costs relative to a corresponding retail price.

Price regulation is normally only applied to operators with significant market power (SMP), since non-SMP players will normally experience the discipline of competition, and price control should not be applied at all in competitive markets. There will be individual cases, however, where even an SMP player's prices are effectively subject to competition, in which case control is no longer relevant and should be withdrawn. In other cases, however, non-SMP players may have effective market power. Two examples of this are call termination and mobile roaming charges, where regulatory intervention to restrain market abuse may be as appropriate as with a monopolist.

It is very important that price regulation should provide **incentive regulation**, that is, give the regulated party an incentive to be efficient and to minimise costs. The price control of baskets of goods as opposed to individual prices gives operators more commercial freedom within the aims and purposes of price control.

4.4.2   Rate-of-return control

Rate-of-return control limits the regulated company's profits to a prescribed return on the assets employed. The permitted return is periodically reviewed and represents an assessment of what would constitute normal profit, were the market to be competitive. Rate-of-return control rests on the simple theory that because suppliers in competitive markets make normal profits, therefore the imposition of a limit simulates the pressure of competition. This theory is so seriously flawed that it is of mainly theoretical interest nowadays.

The greatest objection to rate-of-return control is that, while it might well simulate the end result of competition, it fails utterly to capture the dynamics of competition. Entrepreneurs do not enter markets in order to make normal profit, but to pursue economic profits. They know, of course, that competitive pressure ultimately erodes economic profit, but they enjoy it while it lasts, using the proceeds to innovate in search of the next opportunity for profit and growth. This leads the industry to seek rising consumer satisfaction at increasing efficiency. However, rate-of-return control guarantees a given return on the assets, no matter how inefficiently they are employed, and so presents no incentive to efficiency. Indeed, it creates a clear disincentive since efficiency gains are simply taken away and handed back to customers.

There is some theoretical evidence that rate-of-return control militates against allocative efficiency. This is because it provides an incentive for the controlled company to over-invest, the so-called phenomenon of 'gold-plating', simply to expand the base upon which its allowed profits will be calculated. The economists Averch and Johnson demonstrated that where a company's permitted rate of return exceeds its cost of capital, it would pay it to substitute capital for other factors of production, and thus not to minimise its costs. Further discussion of the Averch-Johnson effect may be found in Reference 2. More prosaically, the controlled company will be tempted to use various types of creative accounting to overstate its asset base.

Rate-of-return control may not be as objective as it seems at first sight, since the calculation of permitted return is sensitive to the methods of asset valuation. There is an inevitable compromise between completely objective historic valuations and the more realistic but correspondingly conjectural modern equivalent valuations. In times of high inflation, rate-of-return control might fail completely.

Rate-of-return control does nothing to ensure consumer price stability, and may indeed magnify instabilities at times where input costs or sales volumes are fluctuating strongly. Rate-of-return control requires intense scrutiny of the entire business operation, which is both expensive and intrusive. It exposes the regulator to danger of capture because of its dependence on the regulated company for the information it needs in order to regulate. Finally, the regulator who very reasonably reserved and exercised a right to judge the validity of counting particular capital items in the asset base might find itself intruding on investment decisions.

4.4.3   Direct price control

Direct price control, first used on a national scale by Oftel in 1984 for the newly privatised BT, avoids many of the problems of rate-of-return control and is now widely adopted. Price controls are typically expressed in terms of a permitted rate of increase per annum, and are frequently keyed to the **consumer prices index** (CPI) or (in the UK) the **retail prices index** (RPI) via a formula such as (CPI − X) per cent. The permitted increase may be negative, i.e. a price decrease; this will be the case when X exceeds the year's rate of inflation. Direct control of prices avoids the main problems of rate-of-return control for three reasons. First, it provides a price, not a profit, guarantee, which is more understandable and beneficial to the consumer. Second, it provides incentive regulation, as it leaves the regulated operator the freedom to improve processes and increase efficiency, while retaining the benefit. Finally, it is much simpler for the regulator to implement.

The principal challenges in direct price control are threefold:

* fixing the permitted level of increase (or decrease);
* deciding the periodicity of review;
* selecting the prices to control.

The setting of the level, the value of X in (CPI − X), is a result of intelligent and informed judgement, bearing in mind the likely efficiency gains the operator can make, the growing capability of technology, and the likely volume of sales. The danger of setting price control parameters for other purposes, perhaps as a result of political capture of the regulatory process, is best avoided. The need to estimate volumes is critical for two reasons. First, the telecommunications services business is a very capital-intensive industry where marginal costs are often minimal and prices derive mainly from recovery of fixed costs. Second, the demand for some services, notably the higher priced long distance and international calls, is elastic and price changes will have significant impact on volume. The regulator and the operator will perform their own forecasts and calculations and may arrive at a negotiated view of the final value of X. This need not be viewed as an arbitrary or a bargained figure, but an objectively agreed confluence of judgements. If X is tightened too far, then the incentive value of price regulation may be expelled. The operator might then perceive the price control as rate-of-return control in disguise, and act accordingly.

The periodicity of review needs to be reasonably long, typically three to five years, to provide the operator with a period free from uncertainty over which it can take medium-term investment decisions. There is a corresponding benefit in the form of stability for the consumer. Too long a period will expose the operator to increasing risks and, as a result, a higher cost of capital. At the end of the period, the regulator should, after consulting with other stakeholders in the industry, join with the operator to look ahead at the likely gains and sales volumes for the next period. An element of rebate or surcharge may be entertained for excess or weak profits in the previous period. This should be done only with great care, however, for example where hindsight has revealed errors in forecast volumes or the rate of inflation (if this was the subject of a forecast). Any attempt to recover excess profits for their own sake would send the signal that there had been the intention of rate-of-return control all along.

It is possible to control all prices individually, or to subject all prices individually to a common formula, although it is better to impose controls on the average of a set of prices, or **price basket**. This leaves operators greater commercial pricing freedom, and lessens to some extent the danger of political or non-economic factors from entering the process. It is possible to control more than one basket. The median residential call bill has been used in the UK as a price basket, and, since 1997, the same limit has been applied to other portions of the residential bill distribution. Subsidiary controls may be imposed within a basket to limit the impact on the consumer of price re-balancing. An example in the UK was a restraint on the rate at which the basic line rental component of the residential bill could be re-balanced within the overall price basket.

4.4.4   Cost-based price control

Cost-based price controls are typically used in the setting of wholesale and interconnect rates. These are chosen on the one hand to encourage competition without on the other hand forcing the dominant operator to subsidise competitors. To operate cost-based price control, the regulator must, in agreement with the SMP operator, determine the cost of providing a service and add a margin to cover the cost of capital. The resulting prices will often depend on a forecast of volume.

The costs of providing a service may be determined in a number of ways:

- marginal costs;
- stand-alone costs;
- fully allocated costs;
- incremental costs;
- long-run incremental costs;
- avoided costs for non-provision.

Marginal costing would produce unfairly low prices in a capital-intensive industry where the marginal cost of one call may be negligible. In many contexts marginal cost pricing would be beyond doubt predatory. **Stand-alone costs** (SAC) are the hypothetical costs that would be incurred by a supplier supplying only the particular service in question, in a competitive market and operating at economic capacity. This figure will normally be higher than the costs of a vertically integrated supplier (unless it has diseconomies of scope), but may be useful in setting a price ceiling. **Fully allocated costs** (FAC) take the costs to the supplier in question, allocating to the particular service a fair proportion of common costs. Fair cost allocation is far from trivial to perform. A problem with this approach is that it may need manipulating to reflect current best practice efficiency as opposed to the supplier's actual cost-efficiency. It should not be allowed to include discretionary central costs such as a plush headquarters building or sports sponsorship. The extent to which defensible

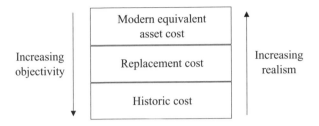

Figure 4.2   Methods of assessing capital costs.

overheads such as research and development, sales, marketing and strategy activities should be included will always excite lively debate.

An **incremental cost** is the difference between the total costs that would be borne by the organisation when not producing the service and when providing it. It differs from simple marginal cost by including the cost of the capital infrastructure needed for the service. This filters out discretionary central costs, and produces a lower result than adjusted fully allocated costs. The **long run incremental cost (LRIC) methodology** is a standard method for determining interconnection rates and includes the long-run service-specific investment that would be necessary to provide the service over time and in volume using the best current technical practice. **Avoided cost** methods estimate the costs that are not incurred when a service is not provided, and might be considered for situations such as a virtual network operation or loop unbundling. The price of the intermediate product delivered to another operator is based on the supplier's end-user retail costs minus the avoided costs of not providing the downstream service.

Any cost estimate will depend on a valuation of the assets deployed to provide the service. The simplest method of valuation is to take the **historic cost**, the price actually paid for the assets, which can be known to the penny. However, with rapid technological development, this is hardly realistic, and the **current replacement cost** obviously provides a better basis. This may be flawed, however, should the best current practice be an altogether different solution, in which case the **modern equivalent asset** (MEA) basis would be the more appropriate. Even this may overstate the value, however, if the new technology brings adjacent opportunities that ought strictly to be filtered out. Unfortunately for regulators and as shown in Figure 4.2, the more realistic the basis of valuation, the greater the elements of judgement in determining it. Fortunately for this industry (and this is far from true for other regulated utility industries), the evolution of technology and rapid network modernisation programmes have led in many cases to a convergence of results from these methods of asset valuation.

4.4.5   Strategies for price control

The details of any system of price control send signals to the market that may affect profoundly the way it develops. The regulator needs to be aware of the consequences of various regimes, and may indeed use them to guide the regulated industry in a chosen strategic direction. A system of price control may engender:

- allocative efficiency for new investment (marginal efficiency);
- efficient utilisation of existing assets (total efficiency);
- dynamic efficiency of future investment (forward-looking efficiency).

The types of price control operated by a national regulatory authority will influence the build versus buy decisions of both dominant and non-dominant operators. Requiring the dominant player to sell wholesale capacity to other operators at the price that would have been obtained in a competitive market encourages the market transition to competition. Forcing a dominant player to sell wholesale capacity to other operators too cheaply may damage the market in various ways.

- It may force the dominant player to make inefficient investments, and reduce its desire to invest where it has choice in the matter.
- It may encourage inefficient players to enter the market.
- It may reduce the incentive for other operators to invest in their own infrastructure.
- Forcing the dominant player to make infrastructure investments that could have been made by others is an act of market management that is economically unsound.
- It may in the long run entrench the dominant position, reducing the growth of competition and perpetuating the dependence of the market on regulation.

Incremental cost-based methods tend to marginal efficiency, while fully allocated cost methods may tend to total efficiency. The use of cost-based wholesale price regulation in conjunction with direct retail price regulation may lead to imbalance between wholesale and retail prices. This might squeeze retail margins, encouraging **inefficient bypass**, where other operators construct their own facilities at possibly higher economic cost than using the incumbent's network. Excessive retail margins will encourage use of incumbent facilities, generating a possible flurry of resale activity that entrenches rather than decreases the dominance of the incumbent. Low regulated prices may starve long-run infrastructure investment while bringing about a deepening dependence on the incumbent. The very existence of regulated incumbent wholesale prices creates a buy incentive for non-SMP players, as the costs of picking up a standard offering at a regulated price may well be below those of negotiating bilateral deals in the free market. An example of this phenomenon exists in the UK mobile market where some operators interconnect with one another via BT's network at regulated prices in preference to using competitive or even direct interconnection.

Requiring an incumbent to provide certain services at regulated prices to all who request them might force it to make investments it would have preferred to defer. In effect, this forces it to grant its competitors an option-to-purchase with a considerable value and cost. This economic option value is not taken into account in the costing methodologies, and may lead to reduced investment by the incumbent. The option value would be small in a non-competitive market, amounting to little more than the time value of money, but rises with the downstream uncertainty that sunken investment could become stranded (that is, made unusable) by regulatory, demand or technological changes.

Table 4.1   Appropriate price controls in different markets.

Market classification	RETAIL MARKET	WHOLESALE MARKET
COMPETITIVE	None	None
PARTLY (or prospectively) COMPETITIVE	Direct, relative to price index	Requirement of non-discrimination
NON-COMPETITIVE	Direct, relative to price index	Direct, relative to price index Cost-based ('cost plus') Avoided costs basis ('retail minus') Direct, relative to price index

Table 4.1 shows the types of price control that are appropriate for different markets. Direct price control relative to the retail price index is the most usual for retail markets. The control of price baskets rather than individual prices provides more commercial freedom for price-controlled operators than would be the case with individual control of separate prices. Cost-based regulation (known as 'cost plus') is normal for wholesale services where the SMP player's dominance is entrenched by control of infrastructure not readily duplicated. Nonetheless, some regulators, including the UK, have introduced control relative to an index of prices in this area as markets become more competitive and to provide a greater element of incentive regulation. Wholesale price control may be based on a corresponding retail price minus the avoided costs of retail supply ('retail minus'). This is perhaps the more appropriate choice in two situations. This first is when the corresponding retail price is competitive. The second is with innovative services, where there is a significant risk that cost-plus regulation would stifle investment or fail to reward risk investment by the dominant player. Where markets are competitive, then the dominant player's charges are wholly avoidable and no price control is necessary. Where markets are prospectively or partly competitive, it may be sufficient to require no more than non-discrimination in wholesale pricing, that is, where the dominant player charges the same to external and internal customers.

4.5   Engineering and technology

4.5.1   Technological neutrality

Regulation should be technology-neutral where similar services provided by different technologies are ultimately part of the same market. The giving of privileged treatment to particular technical solutions may distort the market, and markets will be at their most efficient when companies are unfettered to seek the most cost-effective methods of service delivery. Regulations that are grounded on technology or technical solutions are in great danger of becoming dated and so inappropriate. Unnecessarily prescriptive technical commitment raises the intrusiveness of regulation. Nonetheless, technology cannot be avoided altogether in regulation, since some prescription may be necessary for interoperability and interconnection, and where scarce physical resources are involved. Where *services* are different, then obviously regulation should address the different economics of each market, no matter whether the *technologies* are the same or distinct. It is fallacious to appeal to technical neutrality as a reason for reproducing the regulatory regime for one market in another, for example when arguing that broadband services should be regulated in the same way as voice telephony.

Technology should not feature in rules and regulations where there is no need. A temptation to use technology for defining services or service types arises because it may be easy to do this. If the aim were to regulate the different services or service types differently, then such regulation would be subverted by technological change. It was once possible, for example, to regulate broadcast content by attaching conditions to spectrum licences, since spectrum was a key, scarce, resource without which a broadcaster would have no business. Such regulation would need rewriting in a world of digital broadband delivery by cable and satellite. A regulatory distinction between cellular mobile networks operating in the 900 MHz bands and newer 'personal communication' networks operating in the 1,800 MHz band would be an artificial one, since these services do not differ other than by implementing technology. If the real reason for distinctive regulation had to do with other factors such as later entry or market power, then using technology as a proxy for the distinction could give the rules a short useful life.

4.5.2   The role of engineers

Engineers have a significant role to play in telecommunications regulation. Nonetheless, the engineering discipline does not have the primary roles. Economists and lawyers may take many of these[5]. Engineers are required to contribute both to operational and to strategic decisions. The strategic development of regulation depends upon the input of engineers, since it is possible only to regulate for the possible, and it is necessary to foresee disruptive changes. Many regulatory issues require technical expertise in the conduct of consultations, in co-regulatory discussions and in the drafting of guidelines. Subject areas such as numbering, number portability, technical interfaces, interconnection, spectrum management and loop unbundling require competent technical input, often in considerable depth. Rules, determinations and consents cannot be drafted without awareness of the technical solutions that may already exist or come into being, even though they may well be written in ways that avoid commitment to particular engineering solutions.

Some engineering-intensive activities have declined in importance since the early days of market liberalisation. These include:

- equipment certification and approvals;
- wiring standards and codes of practice;
- network routeing rules.

These are areas where a lot of work was necessary to construct a competitive framework out of the old monopoly. In many cases, problems have now been solved or ongoing aspects handed to the industry by way of co-regulation. One area, however, where the need for technical input remains as strong as ever is that of radio spectrum regulation.

4.6   Notes

1   This rounded calculation is based on two copper conductors with an un-insulated weight of 225 kg/km (800 lb/mile) over a distance of 650 km, with copper priced at $1,625 per tonne, the London Metal Exchange closing price on 11th July 2002.
2   This assumes dense wavelength division multiplexing (DWDM) with a total bandwidth of 40 Gbit/s. This sustains about 600,000 one-way telephone channels at 64 kbit/s or 300 TV channels at 120 Mbit/s. Higher capacities are in prospect as technical development is currently rapid.
3   Cellnet, jointly owned by BT and Securicor at foundation in 1984, was taken into full BT ownership in 1999 and demerged as the mobile operator mmO$_2$ in 2001.
4   The determination that mmO$_2$ (formerly Cellnet) and Vodafone have market influence was withdrawn by Oftel in April 2002 [3].
5   People of engineering background can and do acquire the skills to acquit themselves well in these roles, of course.

4.7   References

1   FLETCHER, M.: 'Cable telephony: coming to the market', *Telecommunications*, September 1997, quoted in Reference 2
2   WHEATLEY, J. J.: 'World Telecommunications Economics' (IEE Books, London, 1999)
3   'Determinations to remove the determination that Vodafone and BT Cellnet have market influence'. Oftel, April 2002

Chapter 5

Interconnection

5.1 Definitions and overview

Interconnection between separately owned telecommunications networks is as old as the industry itself. Even yesterday's monopoly networks needed to interconnect for the purposes of international communications. Various standards-setting bodies laid down the technical interfaces and operational basis, enabling telecommunications companies to operate together with an acceptable and well-understood quality of service. The **International Telecommunications Union** (ITU) is a United Nations charter organisation and a major agent for standardisation. The main concern of this chapter, however, is not with technical standards but with the regulatory arrangements necessary to oversee successful interconnection and interoperation between separate networks in competitive markets.

It is self-evident that separate telecommunications networks must interconnect with others in a competitive market. Accordingly, regulators insist that all public telecommunications operators must be willing to enter into negotiation with any operator that requests interconnection, and that dominant operators must interconnect when requested. The commercial terms for dominant operators usually have to be based on the **long run incremental costs** (LRIC) of providing the interconnection, and national regulatory authorities will determine terms and conditions where operators have been unable to agree them bilaterally.

Interconnection is essential for three reasons: for consumer benefit, to encourage competition and to prevent abuse of market dominance. The consumer interest is best served when a customer of any network is able to make calls or send messages to any compatible terminal, regardless of whichever operator supplies service to that destination. Competition is encouraged when a new entrant's customers can receive traffic from or originate traffic to any other user within a marketplace, and when it can supply network services to a wider population of customers than the limited number who are directly connected as its own customers. Regulation is not normally necessary to compel operators without market power to interconnect, as they will usually find it in their own interests to connect at least with the dominant player. On the other hand, the dominant player, who alone could maintain a viable business while refusing to interconnect, cannot be allowed to do so since this would be an abuse of market power that would be highly effective in excluding competitors from the market.

The principles of regulation for interconnection have been formulated as follows in Reference 1, echoing earlier provisions in Reference 2.

- Dominant operators must provide interconnection to their networks when the demands are reasonable.
- The interconnection conditions must comply with the fundamental principles of fairness, proportionality and non-discrimination.
- Such conditions are to be negotiated by the parties, with intervention by the regulatory authority in the case that they cannot agree.
- Dominant operators' prices must be cost-based with appropriate accounting separation.
- Dominant players must publish a reference interconnection offer, that is make known in advance non-discriminatory prices and terms for their interconnection services.

The vast majority of operators in a competitive market will choose to interconnect with the dominant player, usually the former monopolist. They may also interconnect bilaterally with one another. Since it is unreasonable to expect that all operators will choose to do this with all other networks, the dominant player is normally obliged to provide **transit interconnection** services, providing intermediate linkage between originating and terminating operators, as shown in Figure 5.1.

Interconnecting operators must disclose enough information about their network to allow others to connect with it, for example the network architecture, the location of points of interconnect, the numbering ranges served by each point and the technical interfaces supported at each point of interconnect. Some of this information may be placed in the public domain, while other parts of it may be restricted on a need-to-know basis to other authorised operators. An example of the latter class of data is BT's quarterly publication of network information under the **Network Information Publication Principles** (NIPP).

There are two types of interconnection, distinguished by their functional purpose. The simplest type, **any-to-any interconnection**, applies when one network connects

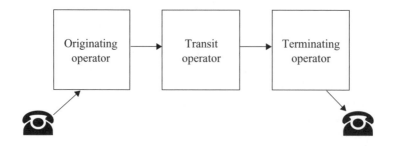

Figure 5.1   Transit and terminating interconnection between operators.

with another in order to complete calls to or from its customers, to obtain transit interconnection with third-party operators, or to construct composite leased circuits. A more complex class of interconnection is **access interconnection**, where an operator with market power is obliged to allow the customers of other operators to gain indirect access to those operators' networks via its ubiquitous access network.

5.2 Physical arrangements for interconnect

5.2.1 Methods of interconnection

A physical connection between two networks, A and B, may have a **Point of Connection** (POC) formed by a network termination point owned by the one operator but in the other's building, or it may have a notional point of connection somewhere in the physical link between the two networks. This latter case is known as **In-Span Interconnect** (ISI), and is shown in Figure 5.2. The POC might in practice occur at an agreed joint-box where there is change of ownership of the physical cable or fibre. In the UK, this is usually a joint-box close to the BT exchange. Figure 5.3 illustrates

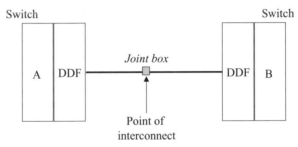

Figure 5.2   In-span interconnect.

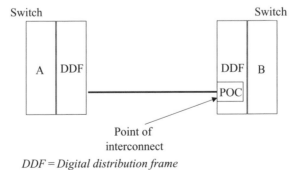

Figure 5.3   In-building interconnect and customer site handover.

POC

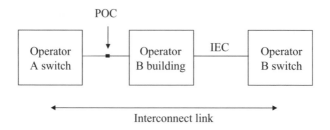

Interconnect link

Figure 5.4   Use of interconnect extension circuit (IEC).

the case where the point of interconnect is in one of the operators' buildings. This is known as **Customer Site Interconnect** (CSI) from the viewpoint of operator A, and **In-Building Interconnect** (IBI) from the point of view of operator B. In the early days, BT used CSI in interconnecting with mobile operators, who had no fixed line plant of their own, and IBI for interconnect circuits with the duopoly competitor, Mercury. ISI has since become the predominant form of interconnection in the UK between BT and operators with infrastructure, while CSI is widely employed in the buildings of operators without networks.

Customer site and in-building interconnect have the obvious problem of requiring one operator to own and to have access to plant in another operator's building. In-building interconnect may have been tolerable for BT under the duopoly with only one competitor, Mercury, but would be impracticable nowadays with many more authorised operators requiring accommodation and personnel access for their interconnections. The interoperability of different manufacturers' Synchronous Digital Hierarchy (SDH) transmission equipment [3] makes in-span interconnection easier to arrange than was the case with the less interoperable Plesiochronous Digital Hierarchy (PDH) systems of the 1980s. In-span interconnect is often the simpler for network management purposes, given that there is only one break of circuit ownership at the point of connection, while CSH and IBI might in practice entail two breaks, one at the in-building terminal point, and a second at the in-span handover of the physical bearer.

A common configuration, shown in Figure 5.4 where an authorised player does not possess sufficient network infrastructure to reach the desired destination, is to use an **Interconnect Extension Circuit** (IEC). This is a leased circuit wholly purchased from and managed by another provider (usually the dominant operator), which extends the connection beyond the directly interconnected buildings to complete an interconnection with another building or switch.

5.2.2   Interconnection and operator network architectures

The particular exchanges at which operators choose to interconnect with other operators' networks may depend on their **network architecture**, or network structure. The incumbent digital networks that replaced strongly hierarchical analogue networks

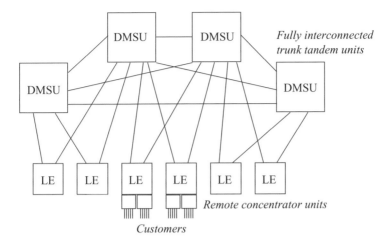

Figure 5.5   Typical incumbent modified hierarchical digital network architecture.

from the 1980s onwards normally have a modified form of hierarchy incorporating multiple parenting of individual local exchange sites as shown in Figure 5.5. Each local exchange has two or more (three in BT's UK network) parents in an upper trunk (or 'transit') switching layer of **Digital Main Switching Units**[1] (DMSUs). Connectivity within the upper (DMSU) layer may follow various patterns.

- Full interconnection of each unit to every other (as in the UK).
- Partial interconnection with some measure of hierarchy.
- Separation of the DMSUs into groups or 'planes' which are fully interconnected within the plane, and where each local exchange has a parent in each plane. A two-plane architecture is employed, for example, in the Netherlands.

    While the purely hierarchical networks typical of the analogue era were primarily designed to concentrate traffic and so minimise the cost of transmission, the modified hierarchy provides much more resilience and may often be able to survive the loss of a link or DMSU without service degradation. Because the minimum economic size for a digital switch is considerably greater than with older analogue switching technology, it is normal for local exchanges in digital networks to be considerably fewer in number than in the predecessor analogue network. The UK has about 750 of these as compared with 6,000 in the analogue network. The remaining local exchange sites contain remote concentrators that are functionally part of one of the consolidated parent local exchanges. The remote concentrators terminate the customers' copper loops.

    New entrant competitor networks display a variety of architectures depending on the operator's size and geographical reach, and may evolve rapidly both with growth and with the reconfigurations that accompany mergers and acquisitions. Three architectures may be identified.

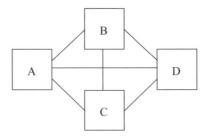

Figure 5.6   Typical new entrant single layer network architecture.

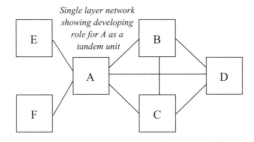

Figure 5.7   New entrant network showing the emergence of a tandem switching layer.

- Simple single layer networks.
- Modified single layer networks where some exchanges have a tandem role and other specialised functions.
- Two-layer networks with a dedicated trunk layer.

A new entrant operator frequently begins with a small number of fully inter-connected local exchanges as illustrated in Figure 5.6, forming a single layered network. Customers, where connected directly, may be 'long-lined' across considerable distances using multiplex systems. A modified single layer architecture, illustrated typically by Figure 5.7, occurs when the exchanges are not fully inter-connected, whether at the outset, or after further exchanges such as E and F have been added. As a result, one or more of the exchanges (A) acquire a role as a tandem switch. Planners usually ensure that each switch has more than one outgoing route as a precaution against link failure. The formal recognition of certain switches as being of a higher layer may occur when their roles as focal points for interconnection with other operators or for inter-regional switching are confirmed. The emergence of a dedicated trunk layer typically follows with network growth. The network becomes easier to manage and dimension when different types of traffic are separated and subject to structured routeing rules. This stage may be hastened by the attractions of a transit layer for hosting so-called **Intelligent Network** (IN) advanced services.

Network operators may choose to interconnect with other operators at some or all of their switches, while two-layer networks may interconnect at their local layer, tandem layer or both layers. Hierarchical network operators often choose to interconnect at their trunk layer, as this may best suit the routeing characteristics of their own network and allows them to concentrate the inter-network accounting software and services there. There is scope for disagreement with interconnecting partners, who might prefer to connect at nearer or otherwise more convenient reception points. Regulatory intervention may be appropriate in individual cases as it is certainly possible for an operator to insist on interconnection points that impose costs on other operators. Interconnecting partners cannot be allowed to place unreasonable demands on one another. The Netherlands requires all operators to nominate an interconnect point for each charge zone (in which they provide service), where they will terminate, at the local interconnect rate, incoming calls to its customers in that zone. Even though a point does not have to be inside its respective geographic zone, the operator could, therefore, draw no direct benefit from perverse choices of connecting location. Limitation of points of interconnect by the dominant player will be subject to the principles of objective need, non-discrimination, and uniformity of practice with respect to its internal use of its network. In practice, issues of network architecture are normally resolved without too much difficulty, and it is the financial aspects of interconnection that often give rise to more serious problems.

When connecting calls from one network to another, originating operators may choose how far they carry the call before handing it over to the terminating operator (or a transit operator). **Near-end handover** and **far-end handover** of a call from originating exchange LA1 in operator A's network to destination exchange LB1 in operator B's network are illustrated in Figure 5.8. Under near-end handover, shown as option N, the originating operator passes the call to the terminating operator at the nearest possible point, and allows the terminating operator to perform any long-distance carriage necessary for the call. Option F, far-end handover, is the reverse, where the originating operating hauls the call as far as possible, to the nearest possible approach to the destination exchange. The interconnection call charge paid by operator A to operator B will be greater in the case of near-end handover, and in an ideal world the rates would be sufficiently cost-orientated to ensure that a resource-optimal choice was made by operator A. In the UK, operators may select any interconnect point, near, far or in-between, and will choose far-end handover if it is more economical to use their own networks than the incumbent's. Near-end handover is unavoidable, however, for calls to non-geographic or mobile numbers, as there is no means of knowing how 'near' the called number is.

5.3   Access interconnection

5.3.1   Methods of access interconnection

A new entrant operator has three options for connecting with its customers.

- It can build its own access networks, and so own the 'last mile' link to the customer.

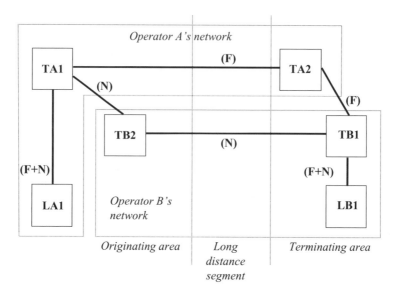

Figure 5.8   Near-end and far-end handover.

- It can rent leased lines from another operator, usually the dominant player, to provide permanent connections between customers and its network.
- It can make use of another operator's switched network to collect traffic from its customers into its own network. This method is known as **indirect access**.

5.3.2   Direct ownership of access network

The construction of copper or fibre access networks by the competitors of the incumbent is a common phenomenon in the downtown areas of major cities. Cable TV operators have constructed combined broadcast TV distribution and telecommunications access networks, taking advantage of economy of scope. Wireless local loop is a potential access technology, while cellular mobile networks provide another access infrastructure.

5.3.3   Leased lines

Where an operator has decided not to invest in physical access plant, customer access may be achieved by the rental of a leased circuit from another operator. The incumbent, dominant operator is the usual source, although other operators may also be able to supply access. It is usual for the significant market power player to be required by regulation to supply leased circuits under a service obligation[2]. Where the circuits are used for the purposes of interconnection, there may be a requirement for the incumbent to supply them under price-regulated terms or on non-discriminatory terms such as would apply for the supplier's own internal or external customers. The resource cost

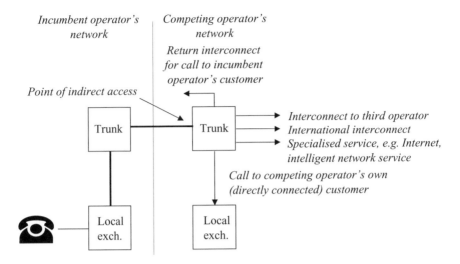

Figure 5.9   Indirect access.

of leased lines, however, makes this solution attractive only for high-usage business customers, leaving indirect access as the fallback method for small business and residential consumers.

5.3.4   Indirect access

A new entrant operator's indirect access customers reach that network using a switched connection through an access provider's network. It follows that the network's customers must also be customers of the access-providing network operator. In the vast majority of cases, the access operator will be the incumbent. Possible connections are shown in Figure 5.9. The indirectly accessed network may, depending on the dialled destination, connect the call to one of its own customers, pass the call on to a third operator, or return the call to the access provider's network. This last may involve an inefficient 'trombone' routeing. The regulatory obligation on an access operator to support indirect access varies from country to country. In the USA, all local exchange carriers (LECs) must provide indirect access to inter-exchange carriers (IXCs). The reason is that the industry is segmented such that an LEC must use the services of an IXC to connect calls beyond its local service area. In Europe and elsewhere, where competing full-facility national and regional operators have been encouraged, only players with significant market power have to support indirect access to their competitors' networks.

Indirect access is a fundamentally complex process, although various techniques have been developed to lessen the burden on the end user. For each call:

- the access network provider must first route the call to a port on the chosen network;

- the caller may have to identify himself or herself to the chosen network for billing purposes, where there is a separate billing relationship. (In the USA, however, this is not necessary as there is no necessary billing relationship, and long-distance call costs are collected by the LECs and passed on in bulk to IXCs);
- the caller must finally transmit the 'dialled number', that is the number of the required destination.

The earliest and crudest indirect access systems exposed the caller explicitly to each of these steps, first with the dialling of an **access code** to signify the network, then after a second dialling tone an account number and a **Personal Identification Number** (PIN) for authentication purposes, and finally the number of the called line. This procedure was extremely burdensome, and was anti-competitive in its effect because it was so much easier to dial the required number directly. The result of this was call carriage in the incumbent (access providing) network. Some relief to the procedure was available at **Private Branch Exchanges** (PBXs) using a software upgrade to insert the access code and PIN. At single lines the network operator might supply its customer with an auto-dialler unit, which uses a single access button to dial the access code and PIN. An example of this was 'Mercury box' used in the UK to gain access to the Mercury network.

5.3.5 'Easy' access

'Easy' access retains the need to dial an access code for indirect access but dispenses with the identification sequence. It exploits the **Calling Line Identification** (CLI) capability of modern exchanges to identify the originating line to the indirectly accessed network. With access codes typically only four digits or less in many countries[3], the disincentive to use indirect access is greatly reduced, especially as the codes are readily programmable in repertory dial terminals and there is no longer a requirement to wait for a second dial tone.

5.3.6 Carrier pre-selection (CPS)

A carrier pre-selection, or non-code access, facility overcomes all the problems of carrier selection and personal identification as far as the end user is concerned. The user makes a default choice of network operator, and the preferred carrier is programmed into the customer 'class-of-service' data at the local exchange. Calls are automatically routed indirectly to the chosen network without user intervention. The default routeing may be over-ridden by the dialling of an explicit access code. The access network provider has, of course, to provide a code for denoting that its own network is chosen for a particular call. The chosen network operator normally sets up CPS by sending an authority, signed by the customer, to the access network provider.

Carrier pre-selection was *de rigeur* in the USA, where it was known as 'Equal access' and where full competition between IXCs was encouraged. Elsewhere, it was felt acceptable to allow an access operator to assume it could use its own network to carry calls not explicitly preceded by a network access code. Nonetheless, CPS is now required under European Directives [4], while other national regulators have inclined to the view that a market providing CPS is a better competitive environment. The UK regulator required operators having significant market power to implement CPS from 2000.

Carrier pre-selection is a complex facility requiring functional specification. The UK service has the following features.

- Users may opt for all calls to be routed to the chosen operator's network. This includes calls to local, national, international, mobile, premium rate, personal, paging, Freephone and other special tariff service numbers.
- Alternatively, users may opt only for national and international calls to be routed to a chosen operator. It is possible to make a different selection of operator for national and international calls.

The UK specification excludes from pre-selection the routeing of certain basic service codes such as the assistance operator, directory enquiries, emergency services and fault reporting services. Calls for unmetered Internet access (see Chapter 9) are also excluded from carrier pre-selection, because of the call volume problems that would arise with the conveyance of unmetered calls across trunk networks.

5.4   Commercial aspects of interconnect

5.4.1   Incremental cost basis for interconnection price control

The regulation of interconnection service prices will depend for a given service on whether it is in competitive supply. The regulator will determine whether this applies to any particular service. With a competitive service, the dominant player's offering is wholly avoidable, other operators will be competing for custom, and price regulation will be unnecessary. Transit services may fall into this category in some markets. Services that are prospectively competitive are ones where some competition exists but it would not be true to say that the dominant player's services were always avoidable. In these cases the regulator may require transparency and non-discrimination between external and internal charges for the services without more explicit price control. Non-competitive interconnection services are charged on a cost-related basis, making provision for reasonable return on the capital. Costs are normally based on the long run incremental cost (LRIC) of provision, and in no case exceeding the stand-alone costs of provision. The regulator's aim is to encourage competition as much as possible without going so far as forcing the dominant player to make indirect subsidy to its competitors. 'Retail minus' or avoided cost pricing is unusual for interconnect, though it is possibly appropriate for newer services with investment risk, where cheaper interconnection might reduce network innovation.

The definition of what constitutes an interconnection service for the purposes of price regulation includes obvious items and others that have evolved as a result of regulatory experience. The obvious examples include indirect access, switched telephony interconnection, and transmission interconnection to enable competitors to make up composite private circuit offerings. Other services may include access to switched data networks, to operator services, to emergency services, to directory

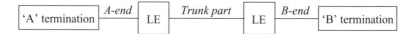

Figure 5.10   Leased circuit retail pricing structure.

enquiry services and to advanced services such as Asynchronous Transfer Multiplex (ATM). The reliance of new entrants on the incumbent's leased line services for access interconnection and for interconnect extension circuits has led in some countries to the classification of these as interconnection services. This relieves new entrants of burdensome retail or retail-minus-avoided-costs charges for these when classed as simple retail private circuits. Experience in this area has led to the concept of **Partial Private Circuits** (PPCs).

5.4.2   Partial Private Circuits (PPCs)

The retail price of a permanent leased circuit may include the following elements:

- the 'A' end access part between the customer premises and the serving local exchange;
- the corresponding 'B' end access part at the other end of the circuit;
- the trunk part, if the 'A' and 'B' ends are served by different local exchanges.

This pricing structure, illustrated in Figure 5.10, is inherent in most network operators' rates for leased circuits, and figured in early interconnection applications. Challenging the high prices involved, interconnecting operators realised that most interconnect leased circuits terminate at a point of interconnect in the incumbent network and so do not, in reality, have a 'B' end. An interconnect extension circuit, when used to connect two of the incumbent's buildings, may not have an 'A' end either. In either case, the above pricing basis was self-evidently not cost-related, a deficiency resolved by the PPC concept which to some extent unbundles the components of a leased circuit. The fine detail of the pricing of PPCs is extremely complex and may change over time. The basic elements of pricing schemes may include:

- the long-run incremental cost of provision;
- a distinction between the parts of the connection that are in competitive supply and those that are not.

5.4.3   Interconnection call charge dynamics

The payment flows that take place when calls are passed between operators are shown in Figure 5.11. It shows the separate cases of 'normal' telephone calls, and calls to special services such as Freephone, number translation services and premium rate services. Some of these, notably premium rate, may include a degree of revenue share, where some of the call revenue is passed onto the owner of the called line as payment for the added value service. In the simple case of ordinary calls, shown using a transit operator, the originating operator pays the transit operator to carry the

Case 1 – ordinary geographic calls

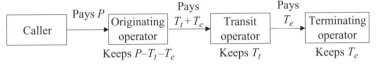

Case 2 – special service calls

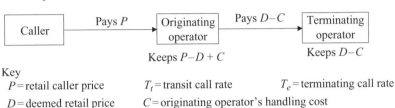

Key
P = retail caller price T_t = transit call rate T_e = terminating call rate
D = deemed retail price C = originating operator's handling cost

Figure 5.11   *Payment flows for interconnected calls.*

call, who in turn pays the terminating operator to complete the call. The prices are metered amounts, by the second or minute. The exact amounts should, at least in an ideal world, reflect the costs borne by the respective operators, so leading to resource optimal-choices of transit operator by the originating operator, and of handover points by both originating and transit operators. These operators are likely to set different rates for different destinations, and the transit price would reflect any requirement to use a second transit operator.

The payment flow for special service calls is conceptually more complex, and is illustrated in Figure 5.11 showing UK practice without the complication of a transit operator. The originating operator charges a retail price P (this and other prices being per minute), giving it a margin over the **deemed retail price** D that the terminating operator will receive. The difference $P - D$ is set by the originating operator (though the regulator may override it), and takes account of the impacts of the call discounts it may offer its customers. The originating operator also retains a handling charge C. This represents the normal incremental network costs of delivering the call to the terminating operator, and an additional **retail uplift**. This retail uplift contains a share of the originating operator's billing costs and some bad debt provision. Cash flows may be negative, as with the Freephone service, where P and D are zero and the called customer pays the terminating operator to receive the calls. In this case, the terminating operator pays the originating operator for bringing the call to its network.

For indirect access, the customer's chosen serving operator is the originating operator for the purposes of Figure 5.11 who must pay a per minute charge to the access network operator for bringing the call to it.

5.4.4 *Interconnection call price structures*

The general requirement that regulated interconnection call rates should recover in total the incremental costs of providing the services leaves considerable freedom for operators (and the regulator if need be) to determine the detailed price structure. Interconnect price structures may be:

- reflective of the corresponding retail structures;
- averaged;
- cost-related.

Retail-reflective interconnect price structures follow the structure, though not of course the actual prices, of the underlying retail rates. Followed in the UK until the mid-1990s for normal call carriage, this method provided for different rates for each service according to two factors:

- time and day (the UK having three tariffing periods of daytime, evening and weekends);
- call carriage distance (the UK using at the time three distance bandings, of local, regional and national calls).

A retail-reflective structure may provide a convenient framework for competing operators who build their business on the difference between the long-distance retail and a local interconnect rate. However, historical retail frameworks are often not closely cost-related, so embedding them in the interconnection regime may inhibit the growth of innovative pricing structures.

Averaged interconnection rates are simple rates that in total recover the allowed incremental costs of the operator collecting them. Such rates apply, for example, in the UK to interconnection rates for special service calls. Their advantage lies in simplicity, though they will distort the market to the extent that they cause rates to diverge from underlying costs. Where there is time averaging, interconnection charges will absorb a greater proportion of operators' revenues in the cheap-rate periods than at other times. Distance averaging may lead to non-optimal use of the incumbent operator's network. This may result in potentially inefficient by-pass for the relatively cheap local call carriage and exploitation on the dearer long-distance segments.

In an attempt to align the call interconnection rate structure closer to underlying costs, BT introduced **Element-Based Charging** (EBC) from the mid-1990s. Recognising that distance transport is nowadays a relatively insignificant factor in network costs, the charging rate between given originating and delivery points is set according as to whether one or two trunk nodes will be traversed by the call. This allows other operators to choose their point of interconnect in a way that reflects economic resource usage of BT's network.

5.4.5   Interconnection accounting

The requirements for the inter-operator accounting and billing of interconnect traffic differ significantly from the functionality needed for consumer and business end-user telecommunications billing. This has led to the development and installation of specially designed **Interconnect Call Accounting** (INCA) systems.

5.5 Technical interfaces for interconnection and interoperability

5.5.1 Basic issues and regulatory approaches

Interoperability of services between interconnected networks and across transit networks implies that services will work regardless of whether they are between customers of the same network or of different networks. It follows that interconnecting networks must deploy harmonised interconnecting interfaces, using and interpreting technical specifications compatibly. Interoperability and interconnection are not synonymous except perhaps for the most basic of services, since it is possible for operators to add to their basic portfolios new services by revising technical interfaces or by exploiting new options within existing interface specifications. Service interoperability is essential for the true development of competition in telecommunications services for the same reasons that interconnect is essential. Interoperability is complicated by technical innovation. Networks will wish to, and should be allowed to, develop innovative services that incorporate an element of differentiation and exclusivity for their customers. This would be a feature of a properly competitive market, and consumers would be the worse off if this process were restrained by slowing development to accommodate the tastes and capabilities of the least efficient operator in the market.

Regulation of technical interfaces in telecommunications presents a complex parallelogram of interacting factors that cannot simply be resolved by ex-ante prescriptions. Regulators must therefore deploy a repertoire of tools blended with case-by-case judgement, balancing the scope for dominance to cause market abuse against the risk that intervention may stifle innovation. A detailed and lengthy discussion of the issues may be found in References 5 and 6. Technical interface regulation must satisfy the following requirements.

- It must ensure that services are interoperable between the customers of different networks.
- It must permit and encourage innovation, where competing operators can improve and differentiate their offerings.
- It must recognise that operators can have control of technical interfaces, and that this can be a source of (or a result of) market power whose abuse should not be permitted.
- It must be aware that control of technical interfaces can result in undue preference, discrimination and exclusion within the market for the supply of telecommunications network equipment.

The weapons within the regulatory armoury are as follows.

- 'Essential interface' provisions.
- Requirements driving operators towards the adoption of open international standards.
- Pre-publication and declaration requirements for all interfaces.
- Use of co-regulatory measures to agree interface issues.
- Ex-post provisions against unfair trading and anti-competitive practice.

5.5.2   *Interoperable service types*

Different services present varying degrees of problem with regard to interoperability. **Co-operative network services** depend critically on network-to-network compatibility. These are services that require originating, terminating and transit networks to operate in concert to support an end-to-end service. They may be defined rigorously as services that would require a functionality specific to the service, in all networks through which a call may pass, to implement them. Basic voice telephony service is an obvious example, although it has presented few practical problems because operators in most liberalised markets were able to inherit established and proven inter-exchange signalling and transmission standards from the former monopoly networks. Newer co-operative network services include:

- Integrated Services Digital Network (ISDN) services;
- Virtual Private Network (VPN) services and Wide Area Centrex;
- Calling Line Identification (CLI);
- Ring-back-when-free, also known as Call Completion to Busy Subscriber (CCBS);
- Number portability (see Chapter 7).

Successful international standardisation efforts have ensured a good level of interoperability of ISDN services in both national and international markets. VPN services, typically more dependent on proprietary signalling standards (or proprietary extensions to international standards), have proved more resistant to inter-network operation. CLI services have achieved successful interoperability in many markets.

Non-co-operative network services include call waiting and other advanced subscriber services like diversion and three-party calls. This type of service does not place any requirements on other networks and its implementation is often confined to the local exchange of the operator offering the service. These services do not, therefore, raise interoperability issues, and individual operators may freely launch these services or not according to their commercial judgment.

Enhanced services are added-value services, such as information services, number translation services and other intelligent network (IN) services. These services rely on underlying, basic co-operative network services to bring the caller and the service platform into connection, but often do not rely on extra network functionality for the caller to enjoy the service itself. Accordingly, enhanced services may not raise interoperability issues.

5.5.3   *Technical interfaces and market power*

An operator with significant market power (SMP) derives that power from the number of customers whose access lines it controls (that is, its market share of the access market), and the ubiquity of its access presence within its market. Accordingly, other operators have no choice but to interconnect with it. Should the SMP operator choose to change its interconnection interfaces, the other players in the market will have no option but to follow. Control over technical interfaces can be abused to distort competition. By controlling technical interface changes and the timing thereof, an

SMP operator can inflict unnecessary costs on other operators, and can unfairly exploit first-mover advantages. Distortions of the market for telecommunications equipment may take place if the SMP operator restricts information about its interfaces to a favoured subset of suppliers.

An SMP operator will be guilty of unfair trading practice regarding interfaces if its choices are sub-optimal and have been designed with the aim of inflating the costs borne by others. The mere causation of costs by interface change, however, is not *per se* an abuse of market power, as innovation will take place in any properly functioning market and players in that market will expect to have to keep up. The costs of interface change may bear differentially between operators, possibly when the larger operator has economies of scale or when technical solutions are optimal for one but not other networks. Even this does not necessarily imply market distortion. It would hardly be of consumer benefit to prevent one operator reaping the benefits of its advantages, simply to protect less efficient players. It requires considerable judgement, therefore, to determine whether a proposed interface change is anti-competitive in intent and effect.

The potential for burdensome interface change is magnified when interfaces are proprietary and lessened when they conform to open international standards. This is because large cost differences between networks may follow when a large operator with an established supplier relationship can evolve proprietary interfaces more cheaply than an operator with limited purchasing power can get its supplier to follow the changes. Where internationally standard solutions are followed, even the purchaser of small quantities of equipment such as network switches obtains the benefits of buying a product with a large market. In the UK, BT's licence requires it to periodically justify its adoption, or perpetuation, of proprietary interconnection interfaces.

Regulators typically have a fourfold strategy for containing unfair interface changes. First, they require pre-notification of changes. This allows for the second stage, of consultation among interested parties. This will reveal the impact of the change as perceived by other players, and allow the substance of any allegations of unfair practice to be assessed. For large changes, a third stage may be the setting up of a co-regulatory committee or working party to plan the execution of the change throughout the market. Fourth, since it is difficult to imagine sufficiently comprehensive ex-ante provisions, ex-post fair trading licence conditions or general competition law will provide the legal background for prosecuting clearly anti-competitive conduct.

Exploitation of first mover advantage takes place if the operator having control over an interface announces a change and then provides a new service for its own customers before competitors have had time to consider and implement (if they so choose) a responsive product. Such action would draw customers towards the network of the dominant player, so enhancing and perpetuating that dominance. A common regulatory approach to this problem is through ex-ante rules requiring pre-publication. This is not, however, a straightforward solution. First, it is difficult to fix the correct and fair term of notice, which in reality will vary case by case and should anyway serve no more than the needs of the most efficient competitors. Second, it is too easy for a proponent of change to comply with the letter of an ex ante rule, while dishonouring it in spirit, either by giving too little useful information to allow others to implement the change, or by making seemingly minor changes later. Judgement and flexibility are critical for regulation, since over prescription may stifle innovation. The dominant player might well decide that it will not proceed at all with a new service should it be compelled to give up first mover advantage and wait until the 'me-too' offerings have arrived. It may try to 'game' the regulator by threatening such a position. The regulator (and other players) may agree pragmatically to accept a limited amount of first-mover exclusivity, and this may be a fair tactic where the dominant player must invest risk capital in developing the new service. Decisions of this type are facilitated where there is a strong and ongoing co-regulatory body reviewing the ongoing strategy for network interfaces and interoperability.

Different problems arise when a non-dominant operator attempts to seize first-mover advantage with a new service requiring interface change. Typically a regulator might view this much more sympathetically than with the SMP operator, given the relative weakness of the proponent's position. To compel the dominant operator to invest in upgrading its interfaces, however, would shift an unfair portion of the commercial risk of the new service onto the dominant player. In this situation a regulator might only compel the dominant operator to support the new service if the underlying functionality were readily available in its network. Alternatively, it could determine a means of equitably sharing the capital investment risk, a course of action that arose over interconnection for Internet access and which is described in Chapter 9.

5.5.4 Regulatory strategy

Regulators typically have the power to mandate the use of certain 'essential interfaces' by both dominant and non-dominant operators. Exercise of this power is intrusive, however, and it is normally held in reserve should the preferable approaches of consultation and co-regulation fail to secure a resolution. The UK regulator Oftel, for example, has never used its essential interface powers, although it has been drawn on a number of occasions into making determinations over the details of interfaces where operators could not agree them.

A sound regulatory strategy for technical interfaces will make extensive use of co-regulation, not just for ad hoc resolution of specific issues, but on an ongoing basis to review the principles taking into account the industry's best joint opinion on technical directions and hence understanding of the upcoming problem areas. The regulator may then decide how, when and where to intervene, bearing in mind the relative risks of stifling innovation and of market abuse taking place, as shown in Table 5.1. It would be appropriate to intervene over interfaces where there was low risk of stifling innovation but high risk of market power abuse. The advanced subscriber services based on the **ISDN User Part** (ISUP) of the telephony and ISDN inter-exchange signalling system [7] are perhaps examples of this class of service. Where technical evolution in the market was rapid and the risks of inhibiting innovation therefore high, as with various **Internet Protocol** (IP) based services, then hands-off regulation where the operators negotiate among themselves may be the more appropriate.

Table 5.1   Regulatory strategy for intervention on technical interfaces.

Risk of stifling innovation by intervention	Risk of market distortion by abuse of market power	Intervention
LOW	LOW	Leave it to the market
LOW	HIGH	Intervene and control
HIGH	LOW	Leave it to the market
HIGH	HIGH	Leave it to the market

5.6   International interconnection accounting

International interconnection of national monopoly networks is a long-established feature of world telecommunications, based technically on the interface recommendations of the **International Telecommunications Union** (ITU), and commercially on the **accounting rate regime** also of the ITU. This system, which is essentially monopolistic in nature, will continue for some time to govern a large portion of the traffic and revenues from international telecommunications. Nonetheless, various forms of legal and illegal bypass have threatened the system, while changes required under the **Basic Telecommunications Agreement** (BTA) of the **World Trade Organisation** (WTO) will over time extend the liberalisation processes seen in many national markets to the world scene. A detailed discussion may be found in Reference 8.

The accounting rate regime is based on bilateral agreements between interconnecting national monopolies, each pair of which agrees an **accounting rate** for telephone calls passing between them. Under the assumption that the corresponding countries have equal stakes in the cable, fibre or satellite link between them, the one pays the other the **settlement rate**, which is half the accounting rate, for carrying the call between the notional mid-point of the link and its network. Periodically, the payments in either direction are balanced and a net settlement made by the country owing the greater amount. Accounting rates have traditionally far exceeded costs, while the richer countries, notably the USA, normally have a strong outward balance of traffic. This allows some countries to subsidise their networks and derive substantial foreign exchange revenues from incoming international telephone calls. The **retail rate** paid by the caller exceeds the accounting rate by a margin chosen by the national operator.

The arrival of competition in some national markets has led to modifications in the accounting rate regime. One of these has been the adoption by some countries of **parallel accounting**, which equalises the accounting rates used by all the operators choosing to link with a particular foreign country. The concept of **proportional return** provides for a national operator having outgoing traffic to complete by right a proportion of incoming calls mirroring its share of the outgoing traffic. Under this principle, some operators found it very worthwhile to offer low prices to capture outgoing international call volume on profitable routes, so gaining a share of their country's incoming calls at accounting rate prices. However, parallel accounting and proportional return, both extensions of the obsolete accounting rate regime, are being displaced by competitive market pricing and are not allowed under the European Union's new (2003) regulatory framework.

The accounting rate regime has been attacked by legal and illegal bypass, and by the threat from the United States in 1997 of unilateral action to replace the accounting rates with benchmark prices. This led to the predictable outrage of some of the benefiting countries. Certain countries' regulations permit bypass, others turn a blind eye despite it being technically illegal, while some have denounced it as akin to smuggling and as theft from the national purse. By-pass may take three forms.

- **International Simple Resale** (ISR) is offered by an operator who installs or rents international capacity, and uses it to capture originated calls in one country and terminate them in another. ISR is possible only when permitted by regulation on both sides.
- **Re-filing** takes place when an operator accepts an incoming call and connects it onwards by means of a fresh call to another destination. Even where illegal, re-filing is hard to detect and so prevent, as the true origin of incoming calls may be unknown to the recipient country.
- **Call-back** is a variation on re-filing where the caller rings a call-back operator to book a call and is phoned in the reverse direction by the call-back operator to complete the call. This usually exploits asymmetry of the bilateral charge rate in the two countries. Call-back is hard to prevent, even where classified as a crime.

The Basic Telecommunications Agreement (BTA) negotiations [9] of the World Trade Organisation will require major suppliers in signatory countries to provide interconnection to foreign operators on cost-oriented, transparent and reasonable terms and in a timely way. They are required to publish a reference interconnection offer. The services must be sufficiently unbundled, so that interconnecting operators do not have to pay for services or network elements they do not need. National Regulatory Authorities must implement these arrangements independently and impartially. The term 'major suppliers' bridges the different terminology and legal concepts of various countries, and denotes operators with dominance, significant market power or control over essential facilities as the case may be. The BTA prescribes any operator's right of access to national services in signatory countries, but does not insist on their right to enter those markets as an operator or seller. This protects the varying needs and public service objectives of different countries, allowing them to liberalise their internal markets at their own speed. The effect of the BTA will be to legalise ISR between corresponding countries and to open the accounting rate system to competitive pressure.

The European Directives oblige SMP players to provide interconnection to international players on transparent, cost-oriented and non-discriminatory terms. Regulators may not insert additional authorisation procedures when an interconnecting partner is foreign, thus opening markets to cross-border service provision.

5.7   Interconnection, service quality and network integrity

All telecommunications transmission links and network components impose some impairment, or loss of quality, on the signals being carried, and the end-to-end quality of services depends heavily on the co-operative design of all parts of the end-to-end path. In the days of monopoly, the one operator having total control of the whole network could design for an assured service quality, while international networking was governed by technical standards agreed by those same monopolies. In competitive markets, satisfying quality of service targets is more complex, partly because the unified control of networks possible under monopoly would be extremely difficult to arrange, but also because markets might seek to satisfy genuine consumer demands for cheaper services at less than traditional standards of quality.

Transmission quality of service may be defined by many parameters: (in the analogue domain) loudness, distortion, stability, delay, echo, quantising distortion, noise, cross-talk; (in the digital domain) delay, jitter, wander, slip, synchronisation, bit error rate. For each of these there is a maximum level for user-acceptability, usually laid down by established technical standards or, if not, by operator policy. To ensure the end-to-end quality of multi-link and cross-network calls, a loss budget is necessary, assigning a permitted degree of impairment to each component that it must not exceed.

Service quality in competitive networks has perhaps not proved the critical issue that some feared. Before liberalisation, pessimists imagined that competition would produce so ill-matched an assortment of network technologies that end-to-end quality would fall disastrously, driving consumers rapidly back into the welcoming arms of the former monopolist. Fortunately, this has not happened in many liberalised markets. Many and perhaps most new entrant operators have simply adopted through their equipment suppliers the same 'standard grade' technology as used by former incumbents, replicating the qualities of voice transmission achieved under monopoly[4]. Digital switching and transmission have helped enormously, since the lower levels of impairment make possible a degree of network-to-network handover of traffic that would have been much more difficult with analogue transmission. Some operators have entered markets with cheap, lower quality services often based on voice compression, which, at least on a single-link basis, provide acceptable or useable quality. Were these to be indiscriminately incorporated into multi-link services, then the feared service quality problems would indeed come to pass. In practice, however, cheaper services have not penetrated mainstream communications, and have been targeted at particular market segments, such as international voice resale and private circuit multiplication.

Regulators normally avoid detailed ex-ante prescriptions about network quality, because this is time-consuming, intrusive, and potentially inhibiting of network technology evolution. They cannot ignore quality altogether, however, and must take measurements because quality variation can be an instrument of discrimination, and could be misused as a way of reducing prices while maintaining margins. Regulators must be vigilant to anticipate and prevent a quality of service debacle in the overall national network of networks, since it would disrupt network integrity and availability to the detriment of all consumers. Furthermore, letting matters drift to the point where 'emergency' solutions were unavoidable might lead to solutions that distorted competition. Regulators typically use co-regulation as the medium of monitoring the market. In the UK, for example, the **Network Interoperability Consultative Committee** (NICC) with its many task groups set up under Oftel, agree interface strategy, design standards such as the **Network Performance Design Standards** (NPDS) and a **Network Code of Practice** (NCOP).

Various proposals have been made for the real-time, automatic management of quality in disparate networks. The key concept is that of a **Running Quality Value** (RQV) parameter or set of parameters to which each network element would add its impairment contribution. If a particular call acquired an unacceptable RQV, the detecting network would reject it. The detailed implementations of such principles are research topics. On the one hand, some may argue that this is the only way to guarantee quality in the increasingly complex patterns of future networks. On the other, it seems hardly acceptable in a competitive market for any one operator's network to have the power unilaterally to decide that another operator's call should not and will not be connected.

Inter-network interfaces are complex and dynamic, and even validated, compliant implementations of a standard interface can threaten the integrity of the interconnected networks. Accordingly, operators and especially incumbents with their large national networks are entitled to require interconnecting operators to undergo interworking testing to ensure network integrity. Regulators should not prevent this, but they must ensure that procedures are not used as a means of obstructing market entry. They must, therefore, contain no requirements beyond those known to be objectively necessary. Key areas of difficulty for switched telephony include: software integrity, feature interaction, grade of service, overload controls and recovery actions. A graded test strategy is often helpful, which might require exhaustive testing to interwork with a completely new switching system but a much less exacting test sequence where the switch and its software build are known and tested already.

5.8   Interface regulation and the ISO 7-layer model

Interface regulation may take interest only in a part and not in the whole of an interface of concern. This is true because interfaces in telecommunications and information technology are often extremely complicated. Each part solves one particular technical problem. While one part may be a cause of contention and difficulty between different operators, other aspects may not be in question by any of the parties concerned. This may be because the uninteresting part has an unchallenged and well-known solution, or, where having a number of possible solutions, it may have little economic import whichever solution is encountered.

The **International Standards Organisation** (ISO) 7-Layer Model [10] provides a conceptual framework, or architecture, for understanding the parts within the totality of an interface between two network nodes or pieces of equipment. A detailed

Table 5.2 ISO Open Systems Interconnection Basic Reference Model.

Layer	Problem solved
APPLICATION LAYER	Performing useful services: • Network terminal emulation • File transfer • Electronic mail • Database access • Other services
PRESENTATION LAYER	Making data understandable to both users: • Data representation and formats • Compression and decompression • Encryption and decryption for security
SESSION LAYER	Managing 'sessions' (or calls, or associations) between two machines: • Dialogue control (one-way or two-way) • Opening and closure of a session • Restart and recovery of a session under fault conditions
TRANSPORT LAYER	Establishing end-to-end communications: • Joining and separating logically separate data flows • Dispatching and reassembling communications in the right order
NETWORK LAYER	Routeing: • Getting data to the right place, passing various network nodes if need be
DATA LINK LAYER	Data transmission: • Error correction • Flow control (that is, sending data only at the speed the receiver can digest it)
PHYSICAL LAYER	Physical connection: • Voltage • Impedance • Timing • Plugs and sockets

description may be found in Reference 11, while a simple introductory guide may be found in Table 5.2. An interface is considered to be composed of a number of layers. Each layer represents a discrete function that may be implemented by hardware or software or a combination of the two, and will be governed by an interface specification for that layer. The function at each layer relies for its operation on the existence and operation of the layer(s) below it. The function at each layer provides services that may be called upon freely by the functions at the layer(s) above it. The purpose of each layer is to solve a problem such that layers above may assume that the problem does not exist, or in other words has been solved.

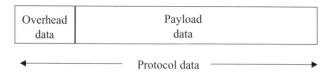

Figure 5.12   Formation of protocol data on an interface or data link.

The separation of an interface into layers allows technology to evolve at one layer, say at the physical layer from copper into faster optical fibre, without precipitating a redesign of the entire interface. Similarly, it may be a higher layer that changes, perhaps because there has been a new version of software, but other layers stay as they are.

Each layer transmits to its corresponding layer at the other end of the link some data, known as **protocol** data. This consists of two parts as shown in Figure 5.12. The **payload data**, or the 'important data', are given it by the higher layers, and this must be sent without change. The layer will take no interest in the contents of the payload data. The layer adds some additional data of its own, known as **overhead data**, which it exchanges with its other end in order to do its job properly. This whole process is recursive, so payload data as seen at one layer will itself consist of overhead data from the next higher layer plus some payload data from the layers above that. It is not unusual for overheads adding up from layer-to-layer to represent a half or more of the data being transmitted. This seems at first sight inefficient, but this need not be the case because the overhead data do useful work and may be unavoidable.

As an example, imagine that the transport layer at exchange A needs to send some data to exchange B. The network layer appends an address as its overhead data; this will see that the data gets to the right place as it travels from exchange to exchange. Between each pair of exchanges, the data link layer adds more overhead in the form of an error detection code so that, should the data arrive corrupted (no physical link is perfect), the other end will realise this, ask for a retransmission and get correct data before letting the network layer route it onwards.

5.9   Notes

1   This is UK terminology. Different names may apply to trunk exchanges elsewhere.
2   European regulation requires significant market power players to supply a minimum set of private circuits. This is not classed as a Universal Service Obligation as there is no requirement for 'affordable' prices, although prices may not be set so high as to have the effect of refusal to supply.
3   Longer codes are now necessary in many markets, for example six digits in the UK.
4   This uniformity does not apply to all quality measures. Others, such as availability and time-to-repair, have been very different.

5.10   References

1 Directive 97/33/EC of the European Parliament and of the Council of 30th June 1997 on interconnection in telecommunications with regard to ensuring universal service and interoperability through application of the principles of Open Network Provision (ONP). The European Parliament, Brussels, 1997

2 Commission Directive 90/388/EC of 28th June 1990 on competition in the markets for telecommunications services. European Commission, Brussels, 1990

3 SEXTON, M. J., and FERGUSON, S. P.: 'Synchronous higher-order digital multiplexing', in FLOOD, J. E., and COCHRANE, P.: 'Transmission Systems' (IEE Books, London, 1991), pp. 243–70

4 Directive 98/61/EC of the European Parliament and of the Council of 24th September 1998 amending Directive 97/33/EC with regard to operator number portability and carrier pre-selection. The European Parliament, Brussels, 1998

5 'Interconnection and interoperability – a framework for competing networks'. Oftel Consultation Paper, 1997

6 'Interconnection and interoperability of services over telephony networks – a statement by the Director General of Telecommunications'. Oftel, April 1998

7 The common channel inter-exchange signalling system No 7 is defined in the ITU-T Q7xx series of recommendations, with the Q76x recommendations referring specifically to the ISDN User Part. ITU, Geneva, published 1988–2002

8 DE VLAAM, H.: 'Cross-border interconnection', Proceedings of the 37th FITCE European Telecommunications Congress, London, 24th–28th August 1998, pp. 37–42

9 'Report of the Group on Basic Telecommunications'. World Trade Organisation S/GBT/4, 15th February 1997

10 'Information technology – Open Systems Interconnection – Basic Reference Model: The Basic Model'. ISO/IEC 7498-1:1994

11 TANENBAUM, A. S.: 'Computer Networks' (Prentice-Hall, Englewood Cliffs, NJ, USA, 1989, 2nd edn.)

Chapter 6

Telecommunications numbering

6.1 Conceptions and misconceptions about numbering

Telecommunications numbers are allocated to individual subscribers and services, and serve as their addresses. They form the method by which a user of telecommunications informs the network of the required destination or service. They are strings of digits, based on the ten digits of the decimal number system. While the range can be extended (and has been extended locally) by the use of additional symbols such as #, *, hexadecimal symbols or letters of the alphabet, any widespread extension is constrained since it would require a major re-organisation of the world telecommunications network.

Telecommunications numbering is a subject for regulation in competitive markets because it is a scarce resource. There is a bounded number of numbers. It has the potential to frustrate competition through wasteful assignment, unfair allocation or simply running out of numbers. Bad management of a numbering system results in numbering volatility (that is, frequent changes to numbers), imposing efficiency penalties on the industry and large costs on users. Numbering relates also to the consumer dimension of regulation, since numbering and dialling plans greatly affect the user-friendliness of a telecommunications network.

Telecommunications numbering can appear a paradoxical subject, since some aspects of it possess different characteristics than may be apparent on first impressions. How can a numbering system, after all a pure product of the human intellect, possibly have limits like a physical scarce resource such as radio spectrum or co-location floor space? This is because a *particular* numbering scheme is unquestionably finite, and its bounds once set acquire solidity through the frictional costs of change that would be incurred to reposition them. Numbering may sometimes seem a boring artefact of bureaucratic minutiae, yet is capable of stimulating emotion, irate public opinion, letters to the press, questions in parliament, and of attracting high public profile.

The superficial simplicity of numbering masks the fact that it needs a big mind to keep it all in mind. It has to satisfy a complex web of parallel and sometimes conflicting requirements of capacity, competitive fairness, long life, smooth migration, user-friendliness, tariff linkage and implementability, notwithstanding that no single one of these is, on its own, beyond the understanding of a competent secondary school pupil. It is not unusual to find someone convinced of the obviousness of something that is actually quite incorrect. Should such a person be in a position of influence, then mistakes can be and indeed have been made. Regulators will receive in the course of consultations a spectrum of opinion that may include aesthetic and emotional angles. A cacophony of viewpoints has caused some people to experience numbering as an esoteric art, a perception not helped by a shortage of people able to articulate the key issues clearly and simply. There is no established reference on numbering; although many general books on telecommunications devote a section to it, most do so in insufficient depth to sustain a serious interest. Some current and historical insights on the subject may be found in Reference 1.

The impact of new technology, notably the World Wide Web and broadband services, raises a valid question of which numbering and addressing framework or frameworks will be most appropriate in future. The final section of this chapter identifies some of the more futuristic issues, while the next sections examine the problems the regulator must address when managing national numbering systems.

6.2   Numbering fundamentals

6.2.1   Growth pressure

All countries have experienced pressure on their numbering systems leading to much reorganisation in recent years. This follows from straightforward demand growth. For most of the 20th century, number demand has tracked the growth in exchange connections (or lines), for example in the UK at an average compound rate of 5.5 per cent per annum from 1.2 million in 1930 to 17.6 million in 1980. After 1980, other factors accelerated numbering demand despite a slowing of main line growth, as follows.

- ISDN lines require multiple addresses at one connection.
- Premium and added-value services sometimes need more than one number per line, such as for distinctive ringing cadences and diversion bypass numbers.
- Direct dialling into multi-line customer installations gives each station an externally presented number in place of the single switchboard number that might previously have served an entire installation.
- Customers may purchase equivalent services from more than one supplier.
- Competition exerts considerable pressure on numbering as new entrants require allocations of capacity.

Additionally, new services such as number translation services ('800' Freephone and such like), personal numbers, information services, radio paging and premium rate services have created fresh demands. Rapid growth of mobile penetration has led to there being more mobile stations than fixed lines in many countries.

All this growth implies longer numbers and hence reorganisations, since a numbering scheme with n significant digits can support at most 10^n numbers, and, as we shall see later, in practice rather less. So, taking the UK city of Leeds as a typical example with an approximate population of 500,000, we find that in 1918 it had five-digit telephone numbers (possibly the first city in the world to have numbers of this length), while six-figure numbering appeared there in 1953 and a seven-digit system made its début in 1995.

6.2.2   Numbering requirements

Telecommunications numbering systems must satisfy a number of requirements. Some of these conflict, calling for balanced judgements.

6.2.2.1   Capacity

A system must at minimum provide enough numbers for the intended lifetime of the system. The British Post Office used to aim for 30-year stability in its forward planning, and was mostly successful until the turbulent 1980s and 1990s. Future proofing against number change is most important, and voices that argue for short-run efficiency within the tightest number length envelopes should not be heeded.

6.2.2.2   Evolutionary potential

Given that any numbering system will eventually need reorganisation to react to foreseen or unforeseen demand change, it must have enough spare capacity to permit reorganisation in a user-friendly way. A random change of numbers (e.g. 200 goes to 31604, 201 goes to 77918 etc.) realised as 'all change at once' would be disastrous, while a logical migration such as 'prefix by 2' with pre- or post-parallel running periods is much better. A parallel running period is one during which people can dial both old and new forms of numbers to secure connection. This eases changeover, letting them schedule the updating of records and reprogramming of stored numbers when they prefer. To permit user-friendly numbering migration, spare capacity must be provided appropriately, for example in the form of spare initial digits or vacant area code space.

6.2.2.3   Accommodation of existing numbering

Most countries would have better numbering systems than they have today, had they been able to turn off their networks and engineer numbering changes from a clean start. In practice, any numbering system must have had a user-friendly migration from the previous system, with the result that earlier systems leave their footprints as numbering evolves.

6.2.2.4   User-friendliness

The user-friendliness of a numbering system depends upon the combined operation of the **numbering plan**, that is the structure and format of numbers, and of the **dialling plan**, the way these numbers are used to make calls. The deployment nowadays of national numbering systems, with either a single national number or a two-part combination of area code plus local number, is largely universal, replacing historic systems based upon exchange names and lists of dialling codes. The detailed design of a national scheme is not an exact science, and subjective trade-offs have to be made.

- Smaller coding areas imply shorter local numbers but greater use of area codes, while bigger areas imply the dialling of more digits for local calls.
- Numbering areas should have cognitive resonance with users and their lifestyles, that is, they should make community sense.
- Any system of numbering area subdivision, or a requirement for numbering areas to track other partitions such as states, counties or tariff zones, may impact the ability of the scheme to utilise its numbering space effectively.
- It is possible to provide more or less user information through the structure of numbers, for example using initial digits to show a zone of the country or a sector of a city. The alternative is 'flat' numbering where number ranges appear to be randomly distributed with the result that geographical information, while possibly there, is harder for the average user to penetrate.

6.2.2.5   Support of competition

In competitive markets, the numbering system should provide enough capacity to allow all the players who want to enter the market to do so without facing discriminatory burdens. It should not create spurious advantages for particular players in the form of shorter numbers, more easily called numbers or otherwise more desirable numbers. It should also support number portability.

6.2.2.6   Tariff implications

Some countries may require numbers to have visible tariff significance, for example by saying all calls within a coding area are local calls, or having premium rate and mobile numbers readily recognisable.

6.2.2.7   Implementability

A numbering system must be realisable by the switches used by all the operators in the network. In the days of electro-mechanical switching, and especially with the step-by-step systems used in the UK and elsewhere, engineering was a major constraint and most numbering scheme design was dominated by implementation considerations. Nowadays, digital stored program controlled (SPC) exchanges can support any reasonable numbering system. It is still necessary, however, to fix rules about the maximum length of numbers, and the number of digits that have to be examined to determine both the charge rate of a call and the routeing, that is the serving operator's destination exchange or trunk node.

6.2.2.8   International dimensions

National numbering systems must conform to international standards to allow international calls to be dialled between countries. This is not normally a problem under the prevailing ITU Recommendations [2, 3]. Countries need to formulate any special requirements that may apply locally. Examples include the integration of neighbouring numbering schemes (such as between Singapore and Malaysia, or the USA, Canada and the Caribbean), or the reservation of numbers and codes in the one country to give cross-border access to all or part of a nearby territory. European directives governing various principles of numbering do not place much constraint on individual country schemes, save for the common international prefix 00 and emergency code 112.

6.2.3 Features of numbering systems

Most countries nowadays have national numbering systems where their telephone numbers contain all that is necessary to make a call, superseding historic systems with city or exchange names and needing dialling code lists. A few countries, for example Norway and Denmark with their eight-digit numbering systems, have unitary national systems where the same digits are dialled regardless of location. France and Italy also have unitary systems, though this is less common for countries of this size. Larger countries generally have **two-part dialling** systems where their numbers are composites of an **area code** denoting the city or locality and a **local number** identifying a particular subscriber. Figure 6.1 illustrates the structure of a typical two-part national number. The dialling plan requires a dialling of the complete number for a general call, but permits the omission of the area code, that is the dialling of the local number only, for calls within the same numbering area. Market research in countries such as the UK has established that users still want abbreviated local or within-city dialling. Often, the local option is permissive rather than mandatory, catering for users who for any reason dial the entire national form of the number despite the local option being available to them. In a growing number of contexts, notably with mobile phones and calling cards, the local option is often not available.

Countries are increasingly adopting **fixed-length numbering systems** as opposed to **variable-length numbering systems**, since the first are undoubtedly friendly for

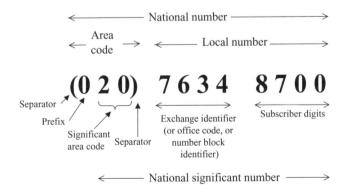

Figure 6.1   Anatomy of a two-part national telephone number.

computers and databases, while users find it comforting to know what a valid number 'looks like'. Many countries (like the UK) have a fixed length numbering system in principle, though with some shorter numbers as retained legacies from the past. Some of these may remain permanently, while many are transitory and will in due course change. Within a fixed length envelope, countries may have **fixed-format** or **multi-format numbers**, for example a short area code with long local numbers in the capital and vice versa up country. The United States has a fixed-length, fixed-format system introduced in 1947 that is now very firmly implanted in people's minds.

Numbering areas in most countries are multi-exchange areas, whose local numbers possess a geographical structure where the early digits of the number (the exchange or number block identifier) indicate a particular locality. The extent to which this structure is exposed to users varies. Historically in the UK and elsewhere, local structuring was very obviously externalised when numbers may have been based on the letters in place names. Increasingly nowadays, the structure of city numbers is not officially exposed, allowing planners greater efficiency through fine-grain allocation of numbers, although observant users still discover geographical correlations with initial digits for themselves.

The numbers in most countries with two-part dialling start with an invariant **prefix** digit, usually zero. It serves to tell a local exchange whether a local or national number is being dialled, so obviously no local numbers can begin with the digit. There are countries with two-part dialling systems but without a prefix in the numbering plan, though in these cases a prefix usually enters via the dialling plan instead. An example is Russia, where callers have to dial 8 before an area code. In the USA callers must insert a prefix 1 for a long distance call, although this dialling plan is complicated. A prefix is sometimes needed for a within-area call without an area code, or not needed for a nearby call with an area code, users having to consult lists of codes and local number ranges to select the correct procedure at a given calling location.

All countries have special numbers, often containing few digits, for giving access to special services. Examples include codes used to access the assistance operator, directory enquiries and the like. The UK introduced the code 999 for access to emergency services in 1937; the equivalent code in the USA is 911 and throughout the European Community 112.

6.2.4   Numbering for new services

From the 1960s, numbers began to appear for new services in addition to their use as simple identifiers for the basic exchange line service. The first major service was probably the USA's '800' called-party-pays Freephone service, introduced in 1969 and taken up in other countries from the 1980s. Later services include radio paging, mobile telephony and personal 'find-me-anywhere' numbers. A common approach to numbering these services has been to find free 'area' codes from the national numbering stock. Many countries copied in varying degrees the '800' pattern from the USA for their Freephone services, for example 0800 in the UK, 1800 in the Republic of Ireland and 180 in Australia. Others adopted different patterns, such as

06 in France and the Netherlands and 0130 in Germany, though the international '800' brand is spreading in countries that did not at first make use of it.

Mobile numbers in most countries have separate area codes, allowing operators to levy distinct and usually high charges when fixed line callers dial mobile phones. The USA and Hong Kong, however, allocated mobile numbers from the same ranges being used for fixed lines, that is, they had city area codes. This had a depressant impact on the growth of mobile communications in the USA, as it was impossible to charge callers a differential rate, forcing mobile users to pay to receive incoming calls. Many countries managed to find readily recognisable blocks of codes for their mobiles, for example Germany's 017x and Australia's 01x, although the UK, struggling to locate spare area codes, mingled mobile and premium rate service codes in amongst the stock of ordinary, geographic area codes. This led to complaints that callers could not be sure of the cost of what they were dialling. The UK's 1995 numbering reorganisation made service type a distinct feature of the national numbering scheme. In this scheme, the initial digit after the prefix denotes the service, for example 01... for ordinary fixed-line numbers, and 07... for mobile numbers.

6.2.5   Numbering for competition

National numbering systems were previously the natural property of the monopoly phone companies, who had to allocate numbers only for their own exchanges and customers. The advent of competition raises the question of what sort of numbers the new entrants' customers should have. There are three options.

- Independent numbering: each company develops its own system.
- Partly integrated numbering: each company acquires a block within the national system, for example an area code or code block.
- Fully integrated numbering: each company acquires a numbering range in each city and for each service on an equal basis with the incumbent.

Independent numbering presents serious problems of user-friendliness, as it breaks the coherence of national numbering systems, needing network names for each serving company, like the 'Leeds National' or 'Leeds PO' exchanges in the UK before 1912. Confusion would be inevitable when the same numbers appeared on two or more networks. It is doubtful whether operators really want differentiated numbering systems, since this factor will affect few customers' decisions to purchase.

Partly integrated numbering gives each network operator a partition of the national numbering system, perhaps an area code or block of area codes. A problem with this procedure is that one must estimate the capacity required, that is, the operator's market share, and then possibly cap it by the amount of capacity allocated. On exhaustion, that operator's customers and most likely others might suffer number changes or dialling plan complications to generate extra capacity. By tying number ranges to suppliers, this solution 'brands' numbers; this is not necessarily a competitive advantage and it certainly militates against number portability. The assignment of area codes to specific operators is therefore unusual for the ordinary fixed line service, although common practice for mobile operators.

Fully integrated numbering gives each operator an allocation of numbers within each coding area where it chooses to provide service. Taking the UK in the 1980s as an example, a Mercury block of 10,000 numbers thus looked just like a BT block of similar size, for example Ipswich 20xxxx or, in London, 01-528 xxxx, and was distinguishable as a Mercury number only by someone taking interest in such detail. This solution, now common for fixed line services, creates a level numbering playing field for all, is consistent with number portability, but does of course amplify the capacity demands on the national numbering scheme. Competition has played its part as a significant and perhaps the largest cause of numbering volatility in recent years.

6.3   Management of numbering

6.3.1   First principles: construction and evolution

The basic construction and evolutionary management of telecommunications numbering systems rest on certain first principles that follow from the need of the system to be deterministic, and to possess the right forms of spare capacity for the system to evolve over time.

A numbering system has to be deterministic if it is to be readily implementable. This has a fundamental impact on telecommunications numbering systems design, forcing them into bounded ranges and preventing the infinity of numbers that would be possible with an intuitive 'count from one' approach. Suppose a user dialled or keyed, say, 5. An exchange could have no way of knowing whether he or she wanted the number 'five', or whether this was merely the start of a longer sequence. This problem recurs regardless of the switching technology, whether it is a mechanical system that must have outlet 5 connected to 'something', or a software finite state analyser that needs to reach a defined state after processing each digit. (There is a potential method with modern exchanges for resolving indeterminacy, which is to wait for a pre-set time or send a 'Have you finished?' prompt to determine the end of dialling. Nonetheless, this is hardly serviceable and finds only niche application.) Fixed length numbering is a common though not necessary solution to making numbers deterministic. All one has to ensure is that whenever a pattern is allocated, then any longer strings having that pattern as a leading sub-string must be occluded by it. So, if 5 were a valid code or number, it would then be impossible also to have 51, 512 or 555 1234. Short numbers have a power to occupy large amounts of numbering space, since an initial digit allocation takes 10 per cent of the capacity, an initial two-digit pair 1 per cent and so on, regardless of number lengths.

Good numbering system managers have the sense to leave spare initial digits. If, say, a four-digit numbering system were to exhaust by reaching 9999, then it would not be possible simply to continue with 10000, as this would conflict with already allocated numbers (e.g. 1000). Instead, it would be necessary to extend the system by transforming it into a five-digit (or longer) system, starting by projecting the existing numbers into the new space. Given a vacant initial digit (say 7), then the old numbers could be projected into the new range by prefixing them with this digit. There could be no conflict of these new 7xxxx numbers with old 7xxx numbers since there were

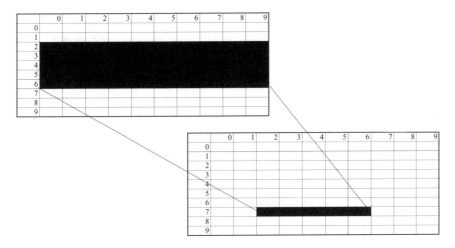

Figure 6.2 Example migration of a numbering scheme by lengthening of numbers.

none. Because exchanges can then easily distinguish old from new numbers, it is possible to have a period of parallel running before and after the nominal change date, and to divert new numbers dialled early or old numbers dialled late to helpful recorded announcements. After a fallowing period (often set at 2 years) for the old numbers to be effectively forgotten, then new capacity such as 3xxxx that would at first have conflicted with old numbers becomes available. Figure 6.2 shows this evolution process diagrammatically. A square grid divides the numbering system into 100 compartments based on initial digit pairs, the vertical axis representing the first digit and the horizontal the second. A four digit numbering system from 2000–6999, thus occupying 50 per cent of the total system capacity and supporting 5,000 numbers, migrates to a five-figure system by prefixing with 7. In the new system, the same 5,000 numbers are supported while occupying 5 per cent of the new capacity.

If a vacant initial digit, sometimes known as an **escape digit**, does not exist for any reason, then migration to a longer range is more difficult, though not impossible. One far from comfortable approach is a 'step-change' where all numbers change at a specified time without parallel running. Because of conflicts, someone dialling a new number before the change may misroute to the wrong person, while someone dialling an old number after the change may dial an incomplete pattern and so get nothing at all. An approach that preserves the parallel running property at the cost of spreading user disruption over much more time, is the 'two-phase' type of change, which exploits a vacant initial two-digit pair. Suppose '38xx' had been vacant in the old system: one would first prefix 8xxx with 3 to make a range 38xxx; after 2 years' fallowing, 8 is now a valid prefix to transform the remaining old numbers.

6.3.2 Capacity: the demand side

The demand for number capacity has accelerated rapidly over the past 20 years in most markets, and extrapolation of past trends is not at all useful for predicting future requirements. Oftel's consultants [4] used a demographic approach to obtain a saturation requirement for the UK of 390 million numbers, and this figure, nearly 7 per head of population, is possibly a useful planning benchmark for other developed markets. Numbers were first classified into ordinary geographic telephone numbers (fixed lines), personal and mobile numbers and finally information services. The potential for ordinary numbers was set at three per worker and two per household (a total of 145 million), whilst the ultimate demand for mobile and personal numbers was estimated at one to each person and three to each worker (also adding up to 145 million). The estimate of 100 million for information services is arbitrary.

One might hope that this planning basis will satisfy the 30-year stabilities achieved in previous generations, although it would be difficult in a time of rapid technology change to be confident of this. Present planning estimates cover only foreseen services, many of which were unthinkable as recently as 30 or 40 years ago. Some think that a higher level of stability is now necessary, and that we have a 'last chance' to get numbering right for the new 21st century. This follows because numbers are often embedded in computer applications, with the result that numbering volatility could provoke unexpected application failures, as people feared for the millennium date change. Nevertheless, estimates could be, and probably will be, defeated by unforeseen innovations, perhaps when we start giving telephone numbers to domestic appliances for telemetry and remote control. Arguably, we need to admit that the only certainty is change, and that the only security, therefore, is preparedness for change.

6.3.3   Capacity: the supply side

A fixed-length numbering system with n significant digits supplies a **basic capacity** of 10^n numbers, though this basic capacity is subject to a variety of loss mechanisms that prevent a surprisingly high portion of this from being usefully available. It is necessary to distinguish **significant digits** from numbering plan digits in countries that have a prefix, as the unvarying prefix does not contribute to capacity, so, for example, the 11-figure numbers in the UK are part of a **10-figure system** with a basic capacity of 10,000 million numbers. Since this is over 170 times the national population and 25 times the saturated demand estimate of 390 million, it is valid to question why this and similar systems need as many digits as they have.

Loss mechanisms reduce the numbering capacity available for use at the three levels of a national numbering system:

• the **national** level, where structural demarcations and area codes are defined;
• the **city** or **area** level, where number blocks are allocated to specific operators' exchanges;
• the **block** or **subscriber** level, where specific numbers are assigned to individual services and users.

These loss mechanisms and the amount of capacity they absorb are shown in Table 6.1. These proportions are only guidelines and may vary significantly for any particular system, since numbering is not an exact science. The mechanisms may be categorised into **primary loss mechanisms** (the first three) which may be very

Table 6.1   Loss mechanisms in numbering capacity supply.

Mechanism	National level	City/area level	Block level	Approx. size
Reserved initial digit(s)	●	●		20%
Escape digits (provision for evolution)	●	●		20%
Planning overshoot provision		●		20%
Churn loss			●	5–10%
Granularity loss		●		Perhaps 50%
Partitioning loss	●			Up to 90% or worse

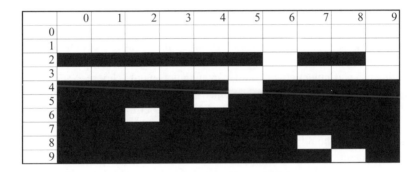

Figure 6.3   Well designed city numbering system at exhaustion.

difficult to relieve, and **secondary loss mechanisms** (the final three) whose impact can be mitigated by careful design and management.

Reserved initial digits are digits with which numbers (or area codes) cannot begin. Found in all systems, reserved digits are typically the prefix 0 and 1, allocated for special codes such as inter-network access, the operator, directory enquiries and the like. At the significant area code level, 0 may be reserved to support 00 as the international dialling prefix.

Escape digits provide as described above the evolution potential of a system. A well planned system may provide two escape digits per first digit range in use, say one general initial digit, and a further specific escape route for each initial digit in use. Figure 6.3 shows an example of a well-planned city numbering system at exhaustion with initial digits 0 and 1 reserved. Escape provision together with reserved initial digits limit the safe operation of a city (or area) numbering system to about 60 per cent utilisation of basic capacity.

A **planning overshoot provision** factor is necessary since numbering change takes time, so, if 60 per cent is the operating limit, then something less than this has to be set as the 'eleventh hour' point to trigger a reorganisation. With growth rates currently in the range of 5–10 per cent per annum and numbering change frequently requiring 2 years' planning and execution, 40 per cent practical utilisation is a sensible planning baseline for numbering capacity planning.

Churn loss arises because few operators reallocate numbers immediately after vacation by the current user, to save annoying calls to a new user of that number. In competitive markets, operators may make this as long as 2 years and more in the hope of offering an old number to a customer who chooses to return after an adventure with another supplier. Churn rates vary from market to market, typically 5–10 per cent per annum for fixed lines and much more for mobile operators. The choice of protected period gives scope for discretion in churn loss provision. Planners need to be wary of double counting since churn losses may be masked by granularity loss at some exchanges.

Granularity loss is a problem when numbers are allocated for any reason in larger quanta than the real requirement. For example, BT (and many operators around the world) used to allocate blocks of numbers to individual exchanges in units of 10,000 lines (four subscriber digits). In BT's case the average exchange size was around half this and many were very much smaller, resulting in wastage of numbering capacity. The precise impact of granularity loss depends on the exchange size distributions prevailing in a particular network, and many planners would choose to add granularity factors to the demand side rather than place a reduction factor on the supply side.

Partitioning loss arises typically in national numbering systems, where there is a partitioned structure such as area coding or service digits. The numbering system will break as a whole should one of the partitions exhaust, effectively gearing available capacity on the most demanding points and reducing available utilisation in the ratio of the peak to mean fill per compartment. The UK suffered and still suffers particularly badly from this since its historic (1958) national telephone numbering scheme gave area code blocks of a million numbers not only to cities such as Newcastle, Leeds and Bristol that challenged the 400,000 (40 per cent) barrier in the 1980s and 1990s, but to many other areas whose average fill was nearer 30,000 numbers per area.

6.3.4   Area coding schemes

The construction and evolution of a national area coding scheme is a crucial facet of a numbering system, affecting both its user-friendliness and utilisation efficiency. Care is needed with this aspect of design, since once established it is very difficult to unwind.

A major design decision is whether to adopt **area-homogeneous** coding areas (that is, roughly equal square kilometres) or **population-homogeneous** areas (that is, roughly equal numbers of numbers). The latter solution, followed in the USA, avoids heavy partitioning losses. The former solution, more common in Europe, permits numbering areas to have a relationship with charging. For example, the Netherlands' numbering areas are co-terminal with charging areas, and the local call tariff applies to a call within its own area or to an adjacent area.

Coding areas should ideally have a community identity that makes sense and to which people intuitively relate. The benefits of two-part dialling are diluted if numbering areas are so small that people have to dial area codes for a high proportion of nearby, everyday calls. Arguably, the tight charge-related codes typical in Europe and inherited from historic systems are becoming too small for modern, mobile life patterns. Larger areas, which may break a traditional charging linkage, are beginning to appear in the UK and elsewhere.

Partitioning loss can be lessened through the use of **multi-format numbering systems**, which allocate longer local numbers and hence more capacity to the larger cities. The Netherlands' nine-figure system, for example, has 0ab – cde fghi numbers in 30 larger towns and cities and 0abc – defghi in the 111 remaining areas. Two formats are probably acceptable, although an over-complex array of different formats militates against user acceptability. Some systems use **code derivation**, such as in the UK for a few rural niches and in Australia before the latest (1996–97) reorganisation, where the codes of one format draw capacity away from a related code of different format. For example, Brisbane had 07 xxx xxxx, while the Gold Coast had 075 xxxxxx and Rockhampton 079 xxxxxx, these preventing Brisbane from having 5... or 9... numbers. If users do not present their area codes properly, derivation can create added confusion over the correct digits for local dialling.

The numerical values of area codes are probably not important, as most users quickly learn to recognise their own and neighbouring area codes whatever their actual value. Most countries have an informal regional structure where early digits denote areas and regions. Germany and Austria have digit-by-digit systems of regional subdivision, where 'near' numbers are geographically near and correspondingly 'distant' numbers are usually far away (but not always since boundaries have to appear somewhere). Modern switching systems use look-up tables for routeing, and are less reliant on regional structures for implementation than may have been the case with mechanical switching. The UK and the USA have apparently random distributions of area code values, in both cases following a historical though now obsolete criterion. The UK's codes were first assigned by letters in place names, hence Aberdeen's 01224 traces back to 0224, at first OAB 4; 'near' codes 01223 and 01225 are for distant places, namely Cambridge (OCA 3) and Bath (OBA 5). The USA ranked their codes in reverse order of time-to-dial with rotary instruments, giving the preferred codes to the more important cities. The finest code, 212, went to New York, while the equal next best, 213 and 312, went to Los Angeles and Chicago.

The UK has chosen to set up major partitions using the first digit after the prefix, the so-called 'S' digit, to indicate basic service types as shown in Table 6.2.

6.3.5 Management of city numbering schemes close to exhaustion

When a city (or local area) numbering system nears an exhaustion point under numbering growth, usually set at 40 per cent of the basic capacity of its n-digit numbering system, three available tactics are number length extension with an escape digit, area splitting and area conjoining. Granularity refinement also may provide relief by unblocking locked-in capacity. Extension of the number length provides long-term capacity relief of a factor of ten per added digit. There is immediate generation of new capacity because reserved initial digits become valid when moved to the second

Table 6.2   UK service digit (S-digit) allocations.

Digit	Service
0	Blocked for international prefix 00
1	Legacy geographic area code system
2	New geographic area code system (for use when old areas exhaust)
3	Reserved for further geographic area code expansion
4, 5, 6	Free for future use and for evolution
7	Find-me-anywhere services: mobile, paging, personal numbers
8	Special charge services up to national trunk charge rate: 0800 and other Freephone, local rate, national trunk call rate
9	Premium charge services costing more than national trunk-call rate

position, for example 1xxxx may have been illegal but 71xxxx is not. After the fallowing period, previously used initial digits become free for redeployment, so releasing the entire capacity generated by the number change. This course of growth may be obstructed if the length of the city's combined area code plus local number is already at the maximum permitted length, or there is a shortage of expansion area code space in the national system.

The solution of the **area split** divides an area previously served by one area code into two (or more) separately coded areas. A two-for-one split doubles available capacity, though only after a fallowing period designed to reduce the nuisance calls that might flow to newly issued numbers in the one area that match long established parallel numbers in the other area. This technique, which was exercised in London in 1990 and is common in the USA, has serious drawbacks, however. It is arguably an inadequate solution for numbering stability, since a doubling of capacity imparts only 14 years' relief at 5 per cent per annum growth and 10 years at the nowadays common 7 per cent. London's area split from 01 to 071/081 in 1990, apparently chosen for its attractiveness as an easier-to-execute 'quick fix' solution, needed further relief in 2000, this time by migration from seven-digit to eight-figure numbers. New area codes can be introduced in **overlay** as in Moscow or New York, that is covering exactly the same area as the previous code and in conjunction with it. This is, however, liable to user confusion and creates a competitive disadvantage for new entrants who will have proportionately more of the new numbers, while the majority of numbers under the older and better known code remain in the hands of incumbent players. When an area is geographically split, as was London into an inner and outer area, it is desirable to find areas both having equal sizes and growth rates yet also possessing community significance; this may in practice be very hard to solve. It is difficult to imagine that the United States is completely comfortable with repeated applications of the area splitting technique, but the pervasive and ingrained popularity of the universal ten-figure number format of the North American Numbering Plan (NANP) of 1947 is a constraint.

Area code conjoining is the joining together of two previously separate areas. Its effectiveness arises by mitigating partitioning loss if it liberates vacant capacity in a lightly loaded area for use in an area under pressure. Usually combined with a number length increase, it may enable the country to move to larger coding areas in better resonance with modern lifestyles. Proposals to conjoin numbering areas can provoke unpredictable emotional reactions, such as:

- *'Disgraceful! We in . . . are proud of our identity. We are not just another satellite of . . .'.*
- *'Wonderful! People will rate this area highly when we have high-tech numbers like the City'.*

Full implementation of local dialling without area codes across a unified area may have to await a fallowing period after the individual areas have obtained their longer numbers, since it is most unlikely that the escape prefixes used for the separate old areas are all valid escape digits in every other area being joined.

Granularity refinement has proved useful in relieving numbering pressure in cities under stress. Adverse side effects are the need to analyse more digits for routeing, and a reduction of the transparent geographic logic in city numbers. In the UK, where numbering capacity is normally assigned to operators in blocks of 10,000 lines, Oftel has introduced 1,000-line granularity of allocation to operators in a handful of places designated as numbering conservation areas, substantially postponing disruptive number changes. Meanwhile, BT has since the 1990s routinely made internal allocations of numbers to individual exchanges in 1,000 and occasionally even smaller blocks. The United States has exploited the IN-screening technology developed for number portability (see Chapter 7) to distribute 10,000-line number ranges over two or more operators' exchange areas, a technique known there as **number pooling**.

6.4   Administration of numbering

6.4.1   Role of the regulator

Telecommunications numbering, once the sole prerogative of monopoly operators, is in most countries controlled and administered by the national regulatory authority. A few variations exist, however. In the United States, numbering was subcontracted in 1996 when the North American Numbering Council (NANC) appointed Lockheed Martin as the North American Numbering Plan Administrator (NANPA), because the former administrator, Bellcore, was considered too closely associated with incumbent local operators. BT at first retained control of UK numbering after the 1984 liberalisation, though they were required to consult with Oftel and others in respect of proposed change and development, and Oftel retained powers of determination [5]. In 1994, Oftel acquired full control of the UK numbering system, in line with practice in other countries [6]. Although the strategic development of numbering is an obvious candidate for regulatory activity because it affects competition and consumer interests, the day-to-day administration is often performed by a regulator only on a 'someone has to do it' basis, and this may, therefore, be subcontracted. Specific number change projects such as the UK's 1995 and 2000 number changes, require concentrated resources and may be conducted by specifically constructed bodies in which the industry and of course the regulator play their parts.

The work of the regulator in respect of numbering includes the following tasks.

- Definition of the national numbering scheme. This will normally be expressed in a widely available document, such as the UK's National Numbering Conventions [7].
- Management of numbering change. The regulator must, in consultation with operators and other relevant interests, monitor the utilisation of the numbering scheme, detect the need for alterations and initiate programmes of change. Regulatory rules generally require the holders of number allocations to make periodic returns about their utilisation of capacity. Some changes may be local reorganisations coming under normal day-to-day management, while larger structural changes may result in revisions to the numbering conventions.
- Strategic development of the numbering scheme. It is helpful for the regulator to have a general strategy, setting a pattern for evolution and for what will happen when areas exhaust. This helps give the national numbering coherence, and lessens the risk of complications resulting from piecemeal reaction to individual pressures.
- Consultation. This is a vital process of numbering development, allowing interested bodies to air their viewpoints. These will include operator interests (including virtual operators, service providers and resellers), and representative user interests (chambers of trade, local and regional government, consumer panels and industry user associations).
- Change management. Changes will often start with a regulatory initiative, although it is desirable that implementation proceed by way of industry co-operation. If relevant, there may be liaison with international or regional bodies, for example the International Telecommunications Union (ITU) or the European Commission (EC).
- Specific issue resolution.
- Number block allocations and day-to-day management.

6.4.2   National numbering scheme definition

The definition of a national numbering scheme should include the following elements.

- Administrative matters: scope, goals and principles.
- Eligibility, conditions, application procedures and granularity for allocations of numbering capacity.
- Requirements for the submission of utilisation data by holders of capacity, and arrangements for day-to-day numbering management including numbering change.
- Numbering scheme structure, including number length and permitted number formats. Major partitions (such as the UK's service digits) and enduring subranges will be identified at this stage, although the data fill (such as specific area codes) might not appear at this level of definition.

- Digit analysis requirements. The relationship between numbers and routeing and charging will be defined, setting out the maximum number of digits which need to be analysed to resolve the tariff of a call, and the maximum number to resolve the routeing of a call to a reception point in an operator's network. Operators may not require other operators to discriminate on more digits than this for traffic routeing purposes, although they have independence to decide their internal network routeing rules.
- Other matters as required (such as, for example in the UK, special service codes, Telex numbering, data network identification codes or DNICs).

6.4.3 Day-to-day administration

A national numbering scheme administrator handles applications for numbering capacity and maintains a register of numbering ranges assigned to operators. This information is normally in the public domain and frequently made available on the Internet. These ranges are 10,000-number blocks in the UK[1], each having a characteristic 0Sabc-de... identity, for example 01473 64xxxx, 020 7634 xxxx. Blocks might be listed as typically having one of the following statuses.

- UNAVAILABLE applies to not yet opened ranges, or ranges sterilised to forestall dialling errors.
- ALLOCATED applies to a range in use by a named operator.
- PROTECTED (with a relevant date) applies to blocks that are temporarily unavailable because they correspond to old number ranges and are in fallow after a re-organisation, or to blocks that have been earmarked for forthcoming number re-organisation.
- RESERVED applies to blocks on which an operator has signified interest but which the operator wishes to keep confidential pending activation.

Regulators generally try to allocate the ranges that operators request, but may have to decline them in the interests of good number management, for example if a request would open a new initial digit in a city, possibly compromising evolution potential. A point of difficulty may arise when an operator wants to have consecutive blocks of numbers but can at first justify the allocation of only one of these. In practice a regulator might try be helpful about this unless the area in question was running short of numbers. A regulator may wish on the other hand to discourage an operator trying to reserve corresponding ranges (e.g. 9x...) in different towns thus contriving to stamp a brand on their numbers. A regulator might protect migration routes (that is, escape digits) in accordance with a long-term strategy to a planned future numbering scheme years in advance of the need to effect that migration, so helping the long-term management of numbering change.

Besides range allocation records, number administrators keep other databases, for example the current list of area codes. Additional minutiae of numbering administration may include the housekeeping of signalling point codes, the registration of prefix values or local routeing numbers for number portability (see Chapter 7), inter-network access codes, and special service codes such as for operator access, fault reporting and directory enquiries. Special access codes may fall under three categories.

- Obligatory codes such as emergency (UK 999, USA 911, European 112) which all operators must provide with that code.
- Controlled codes, such as directory enquiries and inter-network access codes, which operators may not have to provide, but if they do must use that code, and if not cannot permit any alternative use for that code.
- Advisory codes, which operators may at their discretion use for the advised or another purpose.

6.4.4   Case study: 0800 numbers in the United Kingdom

In the late 1990s, the UK numbering system faced imminent exhaustion of its '0800' code space for Freephone numbers. The eventual resolution of this by Oftel with industry consultation provides an example of a specific issue whose resolution required considerable effort.

Exhaustion of 0800 number space had been caused partly by exploding demand, but also because of competitive changes in the marketplace. At first, BT had owned the entire range when they introduced it in 1984 (their competitor, Mercury, opened a rival service using the 0500 area code) and had filled it sparsely with customer requests for memorable patterns such as 20 20 20. Later on, Oftel wanted to make the 0800 brand available to other licensed operators and, having proposed to allocate 10,000 block sub-ranges, found these in short supply since BT had sitting ownership of any block that contained at least one allocated number.

Oftel's first proposal was to extend the current six-figure 0800 numbers to a seven-figure numbering range. This would have given plenty of growth potential, and was compliant with the new ten-figure system introduced in 1995. Further capacity could then be made available by the opening of other 080x codes such as 0808. They consulted the operators, who consented. Customer objections materialised, however, when a user group, the Freephone Users' Group, lobbied the parliamentary Trade and Industry Select Committee. This challenge was a novel method of appealing regulatory decisions, and came as a surprise since the operators had previously asserted that they had already consulted their customers and obtained their approval of the change. The substantive problem was that the users of memorable 0800 numbers were very unwilling to give them up or see them mutate into different numbers. This was understandable, given that they had in some cases built brands around these numbers, and anyway would have faced a daunting job contacting various range-owning operators to assemble an alternative collection of memorable seven-digit numbers. At first it seemed as though an irresistible force, the need for more capacity and competition in the 0800 market, had met an immovable object in the form of real and understandable user inertia.

The compromise, finally formulated in 2000, permitted users who so wished to keep legacy six-figure 0800 patterns[2] while migrating the generality of users and all new allocation to seven digits. At first, and well before 2000, all new six-figure number allocation had been stopped. Next, users were contacted, and a goodly number of these proved willing to give up six- for seven-digit numbers in exchange for a generous period of parallel running. This moderated the number of blocks locked into BT by the presence of an incumbent six-figure number. Finally, the industry agreed exceptionally to refine the granularity of allocation from 10,000 to 1,000 number blocks, thus greatly reducing the proportion of the total range locked by BT. This had an operational consequence as operators' systems now had to examine more digits and contain more data to determine the owning operator for each 0800 number.

6.5   Recent issues and broadband numbering

As the 20th century drew to a close and regulators and numbering administrators had re-oriented national numbering systems for modern challenges of growth and competition, there were and are a number of new topics on the agenda.

- Allocation to non-licensed operators.
- Individual Number Allocation.
- Number trading.
- Broadband numbering and the relationship with the Internet.

Most numbering administrators at first gave numbering capacity only to authorised public telecommunications network operators. Others desirous of numbering capacity, such as virtual network operators, independent service providers and resellers, were required to obtain it in collaboration with a supporting operator. The primary motivation for detaching these secondary allocations from authorised operators is to remove the numbering barrier that might otherwise obstruct a company from changing their underlying supporting operator. Regulators (and European Directives) have therefore broadened the range of organisations that may request allocations of number capacity.

Individual Number Allocation (INA) is a facility for a user of a telecommunications service to select any number (subject to its compliance with numbering conventions and of course to its being free) and require any operator to provide him or her with service on that number. This facility follows as a natural counterpart of number portability (see Chapter 7) and is supported for Freephone number services in Germany and the USA.

Number trading refers to the establishment of a framework by which desirable or coveted numbers such as 222222, an obvious scarce resource, may have some market-based mechanism of allocation. Many regulators are currently engaged in consultation processes on the principles and processes that may apply to number trading [8]. Important requirements are the openness and transparency of any trading floor that develops, with bidding for contested numbers. Regulatory measures may be necessary to prevent hoarding and speculative purchase, as these could give rise to economic inefficiency and market failure.

The extent to which so-called broadband and digital age services should have and will require numbers of the traditional E164 type [2] is an open question. Not all of these today make use of the E164 telephony numbering environment for addressing, nor is it in any way obvious that they should. Those whose service dynamics are similar to those of telephony, notably ISDN, do make use of telephone numbers. Users of other services, such as packet data services and Internet access services via Internet Service Providers (ISPs), may use a telephone number to obtain connection with a server, but employ some other addressing scheme serially thereafter to obtain the desired connection or service. A factor possibly militating against the adoption of standard E164 numbers for broadband and advanced data services is the severe capacity shortage manifest in the North American numbering system. The USA, possibly a leader for many broadband age services, may be disinclined to launch services in ways that are consumptive of scarce numbering capacity.

The following addressing schemes are in use for broadband and data services.

- E164 telephone numbers.
- Physical addresses, for example for TV channels.
- Physical addresses obtained via a name server, for example digital TV channel location through an electronic programme navigator.
- Internet addressing (IP addresses).
- Internet addressing via name servers using the domain name system.
- Closed user group addressing, for example virtual circuit numbers for a switched virtual circuit environment.
- Portal addressing, via menus or links provided by a portal owner or service provider.
- Proprietary addressing environments, for example within computer networks.

Services that typically need dedicated terminals and traditionally have not used E164 telephone numbers to access them, such as Internet pages and broadcast television, are unlikely to start using them. Factors that might drive the adoption of E164 telephony numbers as the address 'handles' for people and services are as follows.

- Any service accessing or being accessed from terminals in the telephony domain, for example an ordinary telephone, data modem, mobile phone, will probably have an E164 number.
- Services likely to involve unified access methods by a variety of terminal types that include ordinary and mobile telephones are likely to have E164 numbers. Examples are unified messaging and personal numbering services.
- A broadband service having an imaged or adapted form that can be accessed from ordinary, non-broadband terminals, is likely to have a E164 number presence.

A key aspect of the convergence of the Internet and telephony worlds will be the means by which resources in the one may be accessed using the numbering and naming system of the other. It is possible to envelop one type of identifier within another protocol, for example telephone numbers may be included as part of e-mail address for sending e-mail to the Short Messaging Service (SMS) mailbox of a mobile GSM user. However, the simplicity of basic telephony terminals gives some considerable bias to the use of E164 numbers for accessing Internet resources from these terminals. An alternative for simple terminals, especially where a display is available, is to generate more complex addresses automatically in a network server that interacts with the user by menu selection. User selection may employ speech recognition or numeric keypad buttons.

The attractiveness of e-mail addresses based on the domain system, for example john.smith@xxx.com where 'xxx' is a company domain name, has led some to speculate whether Internet addressing might in time displace the E164 numerical addressing environment by virtue of its apparently greater user-friendliness. However, a little reflection will convince that naming at this level of beauty is hardly scaleable, and there remains the problem of inputting domain names at simple telephony terminals.

A protocol known as ENUM [9] deals with the process of accessing resources connected with a telephone number from the Internet environment. An **Internet Engineering Task Force** (IETF) working group, the Telephone Number Mapping Working Group, has developed ENUM. This is a format for presenting an E164 international telephone number in a format resembling an Internet domain name. This can then be input to the Internet **domain name server** (DNS) system, for retrieval of other **universal resource locators** (URLs) associated with that number (or its owner).

6.6   Notes

1   In a small number of designated conservation areas, a 1,000-block granularity now applies.
2   The settlement allowed users to preserve legacy 0500 numbers also.

6.7   References

1   BUCKLEY, J. F.: 'Telecommunications numbering', *IEE Electronics and Communications Engineering Journal*, 1994, **6**, (3), pp. 119–30
2   'The international public telecommunication numbering plan'. ITU-T Recommendation E164 (ITU-T, Geneva, 1997)
3   'Timetable for the Co-ordinated Implementation of the Full Capacity of the Numbering Plan for the ISDN Era'. ITU-T Recommendation E165 (ITU-T, Geneva, 1988)
4   'Numbering for Telephony Services into the 21st Century'. Oftel consultative document, July 1989
5   BT Operating Licence condition 34 (HMSO, London, 1984)
6   BT Operating Licence condition 34B (HMSO, London, 1994)
7   The UK National Numbering Conventions may be obtained via Oftel's website, *http://www.oftel.gov.uk*
8   'Developing Numbering Administration and Freephone Numbering'. Oftel, May 1999
9   For a summary of ENUM, see the ITU's website at *http://www.itu.int/osg/spu/enum/*

Chapter 7

Number portability

7.1 Introduction

7.1.1 Types of number portability

The portability of a telecommunications number may be defined as the possibility for the user of that service to retain it after some change to the service. Different classes of number portability refer to different changes through which the number is conserved, for example, a change of the user's physical location, a change in the kind of service, or a change of service provider. Service provider portability is an important factor helping the development of a competitive telecommunications market. Without it many domestic and business customers would encounter a real barrier in moving to another supplier, because of reluctance to accept a change of number and so to have to inform their contacts. Regulators therefore use their powers to establish service provider portability in their national telecommunications markets.

Great care is necessary over the terminology for number portability, because terms such as 'number portability', 'geographical portability' or 'call forwarding' tend to acquire more specific meanings in this context than are intuitively implied by the words themselves. Different countries may choose different words, so it is important first to understand the concepts, and then to identify the terms being used in any particular place.

Location portability, sometimes known as **geographic mobility**, is number portability with respect to a physical relocation of the service. This has been available for simple telephone service in most countries for many years, so long as the customer remains with the same provider and stays within the same serving local exchange area. The move takes place by the reconnection at the main distribution frame of the exchange port from the old outgoing external circuit to the new. The term 'geographic portability' should *not* be used for this, as it has acquired a more specific meaning as a particular class of service provider portability.

Location portability beyond the boundaries of the local exchange serving area presents a technical problem, typically solved by network call diversion or with leased lines to provide an out-of-area service. Providers offering location portability beyond the local exchange area may distinguish for pricing and availability purposes between:

- portability within a city or tariffing area such as a UK charge group;
- portability within the area covered by one area code;
- nationwide portability;
- international portability.

Regulators normally regard location portability as an added value service whose provision and cost is left as a matter for the commercial judgment of the supplier. Intervention may take place when there is feature interaction, for example with combined service provider and location porting of a number, and to protect callers from unexpected charges. A caller should pay the rate for the destination as it appears from the number dialled.

Service portability is the ability of a telecommunications user to retain a number across a change of service, for example from an ordinary fixed telephone to a mobile, or to a number translation service such as Freephone. Service portability has a low profile in many markets and may not be commonly available save for closely related services such as analogue telephony to ISDN fixed service. An overriding principle must be that a caller pays according to the number dialled, any excess costs arising from the new service category being borne elsewhere. At least two UK operators offer forms of service portability that allow a customer to front-end a service with a number of a different type (e.g. a mobile having an 'ordinary' number), these being value added services whose customers pay a premium subscription and possibly a carriage cost for received calls. In the United States, where mobile numbers are frequently drawn from the same city number ranges as fixed lines, number portability from a fixed to a mobile service may be allowed as a normal case of service provider portability, known there as **Local Number Portability** (LNP).

Service provider portability refers to the ability of a customer to retain his or her number after changing to a different supplier for the same or a similar service at (in the case of fixed line services) the same location. Many regulators require this form of number portability in their national telecommunications markets, and may mandate it either as a universal requirement on all authorised operators, or as a bilateral obligation for the one operator to provide with another when so requested by the other.

Personal numbering is no substitute for number portability. Personal numbering is a value-added service whose customer purchases a 'find-me-anywhere' number to track short-term changes of location by routeing a call to his or her home, desk, mobile, voice-mail, answering service or whatever. The possibility of subscribing to such a service does not satisfy the basic requirement for the long-term porting of the current number of a basic service. Indeed, for a competitive market to flourish, personal numbers themselves ought to be subject to service provider portability.

7.1.2 Service provider portability

Service provider portability is commonly, though strictly incorrectly, often known simply as 'number portability', and this convention will be assumed from now when the term is not further qualified. It is normally distinguished into three classes by service type, because the technical implementation of service provider portability for the different service types presents different challenges. Regulators may set different priorities on the three types of portability based on the perceived cost-benefits and levels of competition, placing for example fixed line portability at a higher importance than for mobile services. The three classes are as follows.

Geographic Number Portability (GNP) applies to so-called 'ordinary' fixed-line telephone numbers, that is numbers having area codes or prefixes that give them a geographic locality.

Non-Geographic Number Portability (NGNP) applies to added-value service numbers, such as Freephone ('800' type service), number translation services, premium rate numbers and, in principle, personal numbers though not mobile numbers. These services do not normally have intrinsic geographic locality, and are usually implemented using an intelligent network database access at the time of each call.

Mobile Number Portability (MNP) applies to the numbers of cellular mobile and radio-paging users.

7.1.3 Ownership of numbers

Number portability raises and brings into prominence certain philosophical and practical issues about who may be said to 'own' a telephone number or communications address. Numbers are normally seen as belonging to authorised telecommunications operators by virtue of their having allocations of number ranges under the National Numbering Conventions. These numbers are *not* owned inasmuch as the regulator has powers to change numbers when managing the National Numbering Scheme, but they *are* owned in the 'leasehold' sense that the regulator cannot withdraw allocations other than for certain clearly defined breaches of rules.

Number portability raises the question of whether numbers should be regarded more as 'owned' by users of numbers rather than by operators. This viewpoint would allow a telecommunications user to approach an operator, and ask it to provide service on any number of his or her choice, subject to the National Numbering Conventions and of course to its not already being in use by another party. This is known as Individual Number Allocation (INA), and is especially though not exclusively relevant to non-geographic services. It would save the need for a user, having identified a desirable number, from having first to contract with that number's owning operator before porting to the desired operator. INA and the associated topic of number trading are in active discussion in the UK and elsewhere, although in most places the necessary legal, economic, technical and administrative frameworks remain at an early stage of development. INA for Freephone numbers is supported in the USA and in Germany.

7.2   The costs and the benefits of number portability

Service provider number portability is far from trivial to provide and inflicts costs on the industry. This follows because number portability cuts across the normal sequence of call processing in networks. Telecommunications networks typically route calls by sequential analysis of the digits of a telephone number, first the area code to determine a locality (or trunk switching destination), then early digits of the number to determine the serving operator and again the local exchange, and finally the last digits to identify the wanted customer or line. Each number, therefore, has a natural 'home'. Ported numbers must be identified and then handled exceptionally, requiring number screening, re-routeing and forwarding processes. The costs include:

- set-up costs for the technical and management systems to implement portability;
- modification costs for billing, customer care and operational management systems that use telephone numbers as access keys;
- transaction costs for the act of porting a number;
- running costs for the systems to detect and re-route ported numbers, and, depending on the chosen technical solution, possible per call carriage costs resulting from inefficient routeing of calls to ported numbers.

Number portability, however, stimulates competition and so provides benefits for consumers and the industry generally. The number of customers addressable by a new entrant in the fixed telephony market may be expected to increase substantially when number portability is available. The main benefits are as follows.

- Portability causes more customers to migrate to more efficient suppliers than would have done so without it, so lowering the average costs of the industry.
- Customers who would have been willing to change supplier without portability avoid many of the costs of a number change, such as notifying contacts, reprinting stationery and advertising, and repainting vans.
- Callers to people who would have accepted a number change to change supplier without portability save the costs of misdialled calls, directory enquiries, updating address lists and reprogramming repertory diallers, and avoid the disutility of losing contact altogether.
- Greater innovation and efficiency is stimulated in the industry by enhanced competition.

Oftel's consultants demonstrated a net benefit to the UK economy over 10 years of £1.4 billion ($2.0 billion) for geographic number portability [1], and in 1997 showed £98 million ($140 million) for mobile number portability [2]. Though the amounts in these and similar studies elsewhere are modest compared with the overall size of the industry, they convinced policy-makers that number portability was a positive and worthwhile development.

The calculation of these costs and benefits is far from simple. It is important not to indulge in double counting of amounts that are actually only transferred charges, for example an arrangement fee for number portability is a transferred cost from supplier to customer. Number portability may entail the purchase of intelligent network technology that was going to be acquired in whole or in part for other purposes, and may indeed act as a stimulant for the procurement of platforms for new services, such as location portability, number pooling, Individual Number Allocation and other intelligent network services.

7.3   A functional specification for number portability

7.3.1   Basic entities and rules

Number portability is complicated. An essential pre-requisite is a detailed functional specification of the facility, and of the processes that operators will have to deploy both to set up the portability facility and to port individual numbers.

Number porting takes place when a number belonging to one operator's network is transferred to serve as a number on another operator's network. The number thus moved is described as a **ported number**. The operator who owns the number block of which the ported number is a member is known as the **donor operator**, while its network is the **donor network** and the exchange normally giving service to that number range is the **donor exchange**. The operator who takes over service of the ported number is known as the **recipient operator**; service is given to that number from the **recipient exchange**, which is a part of the **recipient network**. The ported number is an **out-ported number** with respect to the donor network, and an **in-ported number** with respect to the recipient network.

An in-ported number must behave in all respects as though it were the true number of the customer to whom it gives service. Besides being obviously the destination of a call to that number, it must also serve as that customer's number for billing and fault reporting, and be transmitted with outgoing calls as the calling line identity. The recipient operator must recognise the number as belonging to it with full responsibility for fault repair, service management and customer care, while likewise the donor operator must recognise that it no longer has the number and carries no customer relationship or service responsibilities for it.

A number once ported can be ported again, this being known as **subsequent porting**. The simplest case of a subsequent port would be when the customer returns to the donor operator. The number then reverts to its original home on the donor network and ceases to be a ported number. If the port is to a new recipient operator, the original donor operator remains the donor operator for that number, while the second recipient operator becomes the recipient operator for that number. The old recipient operator ceases to have a stake in the number save to the extent that it is an agent during the transaction of subsequent porting.

If a customer of a ported number terminates the service provided on that number, then the recipient operator must notify the donor operator, to whom the number reverts. The recipient operator may not retain a released ported number for reallocation to another customer. If the donor operator initiates a numbering change on all the numbers in the block containing a ported number, then the recipient operator must implement that change to the ported number, also honouring any periods of parallel running of old and new numbers that the donor operator may apply to the change.

UK regulations provide for a recipient operator, on finding that it has in-ported a majority of the numbers within a given block, to request reallocation of the block to it, so inverting the ported status of its in-ported numbers into own numbers, while the donor's customers' numbers become out-ported from the new block owner. There are potential problems with this approach, lest the donor's customers should lose end-to-end facilities that do not work across the routeing paths that may be formed after their numbers become in-ported numbers.

A porting functional specification must state the types of numbers that can be ported, normally that numbers may be ported only between like services (geographic, non-geographic or mobile) or between similar services, such as PSTN to ISDN or analogue mobile to digital mobile. In the case of geographic (though not mobile and non-geographic) portability, specifications frequently state that a number can be ported only to an equivalent service at the same physical location.

For specially tariffed services, porting should not significantly increase nor decrease the cost to a caller as a result of the porting. Porting of premium rate numbers must not allow service providers to bypass access codes of practice or parental controls thereto, for example for 'Adult' services or other services where customers must explicitly opt to have access.

7.3.2   Call routeing to ported numbers

Calls to a ported number must of necessity start in an **originating network** and pass to the recipient network, except in the special case where the originating and recipient networks are the same. Depending on the technical implementation and the call routeing, a call may also pass through the donor network and one or more transit networks. Number portability is a co-operative service between operators, so it is essential to specify routeing rules and methods of addressing in and between the networks to ensure that a call reaches the correct recipient exchange. These rules may presuppose a particular technical implementation, of which more below, or a spectrum of solutions from which operators may make a choice.

7.3.3   Number portability and service areas

Many rules (e.g. in the UK and the USA) restrict number portability to a change of supplier at the same physical location. They may, however, permit **subsequent location portability**, that is relocation after the act of porting. Restrictions are necessary since different operators normally have different serving exchange area boundaries. Should a recipient operator apply location portability to an in-ported number so as to move it outside the donor operator's serving exchange area, it could make it very difficult for the donor operator to take back the number, in effect forcing the donor network to provide location portability.

The UK has the concept of the **service area**, an area defined by the donor operator for each of its number ranges and beyond which a recipient operator may not provide location portability with an in-ported number from that donor. This has proved restrictive for BT's competitors, most of whom have larger service areas than BT's

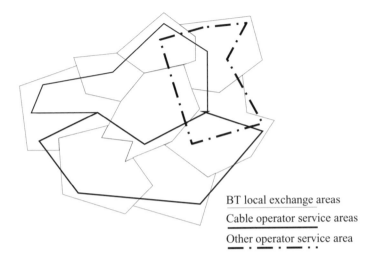

BT local exchange areas

Cable operator service areas

Other operator service area

Figure 7.1   Operator service areas.

6,000 or so historic local exchange areas, as illustrated in Figure 7.1. The service areas of BT's competitors can prove in some circumstances even more limiting than this. Different operators' areas are not the same, so there is a possibility that in some places a number port between them could trap the ported number within a small intersection of areas. A possible though not ideal solution might be to permit location portability outside the donor service area provided the customer knowingly and formally acknowledged that he or she would thenceforth lose the facility of subsequent service provider portability.

The Netherlands has location mobility within the boundary of an area code (Dutch numbering areas are also local call tariffing areas), removing the need for restrictions about donors' service areas. Germany does not encounter a subsequent location mobility problem because numbering areas for all operators follow the same boundaries as incumbent Deutsche Telekom's exchange area boundaries. German service areas are thus co-terminal and do not impact number portability.

7.3.4   Operational processes

The operational processes needed to give effect to number portability are critical to its success, since any friction and delay, whether caused by poorly designed procedures or obstructive attitudes, could damage customer and operator confidence and so impede the effectiveness of portability. Co-regulatory working groups involving the network players should ideally design the procedures, with the regulator facilitating and perhaps providing the chairperson. The operators then have a stake in and ownership of the output, and have less opportunity for obstruction than if the regulator were doing all the work. Ideally the procedures should place minimum inconvenience on the porting customer, while allowing the recipient operator, who

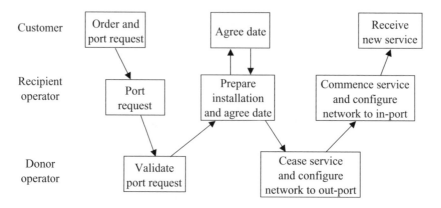

Figure 7.2   Typical flowchart for transfer of service with number portability.

has the greatest incentive to make porting successful, to take overall responsibility for tracking progress. The donor operator must meet agreed service deadlines as defined in the agreed processes, else it would be guilty of anti-competitive behaviour. Typical steps in the porting process for geographic portability are set out as follows and as illustrated in Figure 7.2.

Analogous processes apply to other services, though there are variations such as the delivery of a new handset and SIM card for mobile number porting.

- The customer signs an order for new service and requests number porting.
- The recipient operator requests porting of the number from the donor operator.
- The donor operator checks and validates the request.
- The recipient operator prepares for installation and agrees a porting time with the donor operator. This is normally the point in the process at which the number porting is committed to happen.
- The donor operator ceases service on the customer line and arranges for the network to route calls for the out-ported number to the recipient network.
- The recipient operator connects the customer to its line and arranges for its network to accept in-ported calls to the number.
- The customer receives his or her new service.

There are potential hazards in the process. There is normally a gap between the end of the old service and the start of the new, which should be targeted to a short interval, of an hour or less. Donor validation may raise problems, perhaps if name and address details do not match, or where other donor network services applying at the line may complicate the handover. The customer may fail to meet the appointment giving the recipient operator access to his or her premises to connect the new line. It is helpful to have a fallback procedure to the donor service, and for either network to cover the new subscription with a voicemail service should the gap lengthen for any reason. BT offers an automatic dial-in facility with a security code, allowing the recipient operator to trigger the moment at which porting from BT takes effect. Good relations between junior staff at both operators should be cultivated, as they can do a great deal to hinder or help the process of porting. Apart from the fact that a regulator would view obstruction as anti-competitive behaviour, both operators should understand that they stand to gain from healthy working practices, since number portability is a bilateral benefit.

7.4 Technical implementation of number portability

7.4.1 *General themes: IN and onward routeing solutions*

The challenge for the technical implementation of number portability is to modify the 'normal' way a telephone call is routed, which is by sequential analysis of code and number digits to identify the operator and 'home' local exchange, then the particular line. While regulators may try to avoid mandating a technical implementation in the interests of limiting intrusion, number portability is essentially a co-operative service between networks and it is difficult in practice to avoid some commitment to a technical solution or class of solutions.

There are two basic solutions, namely **onward routeing** methods, and **intelligent network** (IN) approaches. Under onward routeing, all calls are routed in the normal way by digit decode to the 'home' local exchange (that is the donor exchange). This exchange has the job of detecting ported numbers and routeing them onwards to the recipient exchange. With the IN approach, all calls are screened by checking their numbers against a database resource, which identifies ported numbers and their true destination exchanges. This screening may in principle be performed at originating exchanges, or at other transit points in the call path. This is known as the 'IN' approach because it uses the technology of the 'intelligent network', where an exchange recognises that it needs additional resources to connect a call, triggers a data exchange with a separate computer, receiving in return instructions for handling and routeing the call. This same technology is used for special services such as Freephone, premium rate calls and virtual private networks.

The onward routeing and IN approaches have quite different profiles of installation and running costs, and in where costs fall. The principal features of onward routeing methods are as follows.

- All calls require the co-operation of, and have dependence on, the donor network. Under some technical implementations, calls may remain connected through the donor network for the whole duration of the call.
- The cost to implement these solutions is probably at the minimum, as is the time required to set them up (about a year).
- The running costs of these solutions are high because there are carriage costs for calls to ported numbers when they take inefficient routeings via the donor network. Techniques such as drop-back and query-on-release, described later, can mitigate this.
- These methods provide only limited freedom for operators to develop their own technical solutions.

- Operators who do not wish to participate in number portability (and where the regulator gives them this option) can ignore it. All they have to do is route calls to the donor exchange, that is the location apparent from the number.
- Recipient operators' quality of service to customers with in-ported numbers depends to some extent on the performance of the donor network.
- The solution fails completely should the donor network cease operating, for example by reason of bankruptcy. In this case, users of out-ported numbers would lose their service and could only regain it by change of number, even if they had ported their numbers years before.

Contrasting features of solutions based on IN techniques are these.

- All operators in a market must take account of portability. The responsibility for the correct routeing of calls to ported numbers falls on originating operators, not the donor network.
- Calls will, in principle, follow the most efficient routeings to their destinations.
- IN solutions require more outlay of capital and time than do onward routeing solutions. Thereafter, they have potentially lower running costs.
- IN solutions require a database of ported numbers. Someone has to provide and maintain this.
- Operators have a spectrum of technical choices as to where in their networks they perform call screening for portability. They may choose to outsource this responsibility to a transit operator.
- The platforms used for IN solutions may find other applications such as location portability, Individual Number Allocation and number pooling.

7.4.2   Geographic number portability

7.4.2.1   Onward routeing solutions

Figure 7.3 illustrates the flow of a general call to a ported number using onward routeing. The call arrives by number analysis at the donor exchange, from where it is forwarded to the recipient exchange. There are various possibilities for the route taken

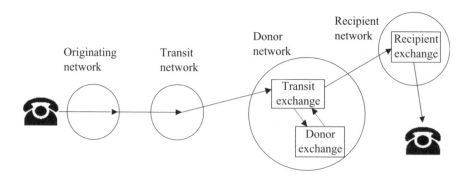

Figure 7.3   Schematic implementation of geographic portability by onward routeing.

by the onward connection, as a 'trombone' back to the donor network transit level (as shown in the diagram), by direct route between donor and recipient exchanges, or via different transit points in the donor, recipient or other networks. In all cases (save when the call originates at the donor exchange), the call takes an inefficient route because of the need to visit the donor exchange.

The onward routeing function at the donor exchange is not the same as the well-known 'call diversion' facility at modern digital exchanges, but is both subtly and profoundly different. For this reason, the terminology 'modified call forwarding' or other names are sometimes applied to denote onward routeing for portability. The standard call diversion facility terminates the incoming call at the subscriber service level in the local exchange, effectively creating a fresh and often separately billable call to the new, diverted target address. The forwarding function for portability handles a call at the *routeing* level, passing it on as a tandem call. Providing onward routeing is thus a facility upgrade to a digital exchange; it is not fair or accurate to argue that the facility is 'already there' because of call diversion. Whilst under the onward routeing solution, the ported number really moves to the recipient exchange, under simple call diversion it would stay at the donor exchange, relying on a 'shadow' number (possible to dial but normally concealed) for the ported customer at the recipient exchange. It is this shadow number that would show as the calling line identity for the ported customer's outgoing calls. Simple call diversion has been used as a crude and early portability solution in the USA, also in Hong Kong and New Zealand. Onward routeing is used in the UK and many other places.

7.4.2.2 Drop-back and query-on-release solutions

Drop-back is one of two techniques available to mitigate the inefficiency of the routeings that can occur with portability based on onward routeing. Under drop-back, illustrated in Figure 7.4, the donor exchange receives the incoming call, and, on determining that it is to a ported number, refuses the call with signalling messages indicating 'porting' and a new destination address. The transit exchange then releases

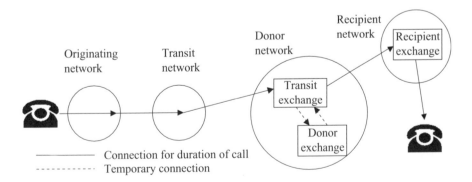

Figure 7.4   Schematic implementation of geographic portability by onward routeing with drop-back.

the link to the donor exchange, and connects the call instead to the recipient network and exchange. This solution depends on the augmentation of the inter-exchange signalling with the new messages and the ported address parameter, and relies on the last exchange before the donor exchange being able to implement the drop-back facility. Consequently, this technique is usually applied within the domain of a single network (for example by BT) and not between different operators' networks. While this solution saves routeing via the donor *exchange*, the connection may still be committed to a routeing via the donor *network*, while processing resources are consumed at the donor exchange for each and every call.

Query-on-release possesses a similarity with drop-back in that the donor exchange reacts to a ported number call with a 'release-on-porting' signalling message. However, whereas with drop-back the donor exchange supplies the destination parameter, under query-on-release the exchange acting on the release uses intelligent network technology to consult a database to obtain the recipient network destination. This solution is thus a hybrid of an IN method to obtain the destination, and a brief dependence on the donor network to determine that the number in question *is* ported. The release-on-porting message could in principle be processed as far back as the originating exchange, but is in practice more likely to be handled within the donor network at the last transit exchange as in Figure 7.4. This method is open to the criticism of being so near to a full IN solution that it is not worth developing separately. It does, however, greatly reduce the database access rate requirement when compared with an IN solution that has to screen all calls.

7.4.2.3   IN solutions and databases

IN, or intelligent network based, solutions rely on the screening of all calls against an external database to determine first that a number has been ported and second to where it has been ported. Screening can in principle be performed in originating exchanges thus permitting optimal call routeing, or at any other exchange in the path to the destination, as shown in Figure 7.5. A visit to the donor network is only necessary should this be the chosen routeing, and for no other reason. The correct delivery of calls to ported numbers is the responsibility of all operators in the market, though the option exists to subcontract this to a transit operator. Screening is in practice often performed at the transit (or trunk) level in two-level networks and usually just before the point of entry into the local network, that is after carriage of the long distance part of a call. It is much more costly to screen at all originating local exchanges.

Any IN solution relies upon the existence of a database of ported numbers. Normally a master database will supply regularly updated copies of the data to individual operators to load into their operational systems. These will typically be highly secure, fault-tolerant, duplicated processing nodes handling tens of thousands of requests per second. The master database may be physically centralised, or it may be distributed with a separate part for each operator. The master database may be provided, maintained and administered by one of the operators, by a non-profit company jointly owned by operators, by a contracted third-party supplier or (unusually) by the government or regulator.

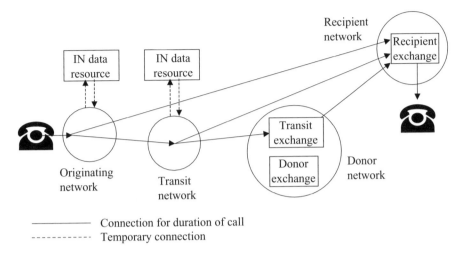

Figure 7.5   Schematic implementation of geographic portability by an IN solution.

7.4.2.4   Addressing for ported calls

Unless screening is to be performed at every single exchange, some mechanism is necessary after detection that a number is ported, to modify the destination number of the call to indicate that it is to be routed to the recipient exchange. With onward routeing methods, the donor exchange inserts the modified number, while under IN and query-on-release solutions it is a database lookup that supplies the modified number. There are three basic forms that a modified number can take.

- No modification solutions, where the externally presented number is used at all times, are applicable only under all-IN approaches such as in the Netherlands, where all networks expect to have to screen numbers.
- **Prefix** solutions add a short prefix to the ported number, to show first that the number is a ported number, and second to identify at least the recipient network (as in Australia, Finland and Germany) and maybe also the recipient exchange (as in the UK). Prefix solutions may be constrained by the overall number length available within the national signalling system, and by the need to ensure that prefixes do not conflict with other valid numbers or facility codes. The UK uses 5xxxxx prefixes that cannot be dialled and so do not interact with other features[1]. A possible management burden with a prefix solution such as in the UK arises when recipient operators make frequent updates to the routeings of incoming calls during network evolution. This inflicts prefix volatility and update burdens on donor operators.
- **Local Routeing Number** (LRN) solutions substitute for the ported number a new number that denotes the recipient operator and exchange, while the ported subscriber number itself is concealed in a supplementary address parameter for reference at the final destination. This, the chosen solution in the USA, uses the

'generic address parameter' in the ITU-T Common Channel Signalling System No 7 (SS7). This parameter is not available for use in Europe, being there reserved for other supplementary services.

7.4.2.5   Routeing rules and signalling

All national networks need to specify certain routeing rules, at minimum to prevent circular routeings and also to set limits on wasteful routeings. In all countries, recipient exchanges must accept and terminate an incoming call to one of their own in-ported numbers, and cannot send it back to the donor network. Many countries, such as the UK, forbid a recipient *exchange* to emit to another network an originating call for one of its own in-ported subscribers, though others do allow this. The Netherlands approach tries to give maximum freedom for operators to make their own trade-off, offsetting their costs of screening against the higher interconnect charges that inefficient routeings will incur. Recipient *networks* may send via the donor network calls that will eventually return to them (though they can save costs by screening against this), and most countries' regulations require donor networks to accept for transit calls to their out-ported numbers. Great care is needed with calls from a donor exchange to one of its out-ported numbers, and many networks have a 'ported' flag in signalling messages to prevent the first transit exchange from sending the call straight back. Temporary circular routeings may be inevitable during the number porting process itself, should the donor network implement the out-porting of a number before the recipient network has updated its data in readiness to in-port it. These are trapped by time-outs or hop counting in the inter-exchange signalling system.

Geographic number portability requires additions to inter-exchange signalling systems. Almost every country has followed a national approach and there is no recognised set of standard modifications to SS7. This matter is receiving attention in the working parties of the International Telecommunications Union.

7.4.3   Non-geographic number portability

Non-geographic services are specially tariffed services that do not have an inherent location, for example Freephone ('800') service, number translation services like 800 but priced at the local rate or national rate, premium rate numbers and personal numbers. Most non-geographic services are implemented as intelligent network services, where the serving operator's or service provider's switch triggers a software application and database reference to determine the destination. Some of these services can be complex and sophisticated, possibly involving an automated caller dialogue or a real-time calculation of the destination depending on time of day, the caller's location or other variables. Many countries allocate blocks of non-geographic numbers to operators and service providers, so that each number has an 'owner' as with geographic numbers, though Germany and the USA now have Individual Number Allocation (INA). Specific numbers may have a high value to their users by being memorable (for example 30 20 10) or having a letter equivalent (such as FLOWERS). The value of some numbers is user-specific, as with, for example,

747747, a number likely to prove attractive to an airline. Given the desirability of particular numbers, non-geographic portability has received a high priority in many markets.

The intelligent network nature of most non-geographic services has made the technical implementation generally less problematic than with geographic portability. Nonetheless, many countries rely essentially on onward routeing solutions. The UK's method initially routed all calls to the donor network, where the IN look-up determined that the number was ported and triggered onward routeing to the recipient operator, where a second IN look-up performed the actual logic of the service. Later versions of the UK solution prevent the trombone routeing when the recipient operator is also the originating operator, by having each operator screen for and trap calls to its own in-ported numbers.

The USA uses a third-party managed database, the Service Management System (SMS), to support portability and Individual Number Allocation for the 800 Freephone service. Transit exchanges examine the database (or a copy of it) to determine the operator who should receive the call, and route it accordingly with no need for a local routeing number or prefix. Germany similarly uses a database, unusually operated by the regulator. Each operator passes non-geographic calls to a transit point able to screen the number, passing it on (or retaining it as the case may be) for the operator or service provider currently supporting that number.

7.4.4 Mobile number portability

Most countries intend in due course to provide mobile number portability (MNP), and early implementers have been the UK, the Netherlands and Hong Kong. Portability of mobile numbers has typically received lower priority than geographic and non-geographic portability, partly because of technical problems but also because mobile markets have been regarded in some countries as more competitive than wireline markets. Most countries have distinct code ranges for mobile numbers, making them recognisable and allowing operators to levy higher charges on calls to mobile numbers. In contrast, Hong Kong and the USA take mobile numbers from their stocks of geographic, city number blocks. On the one hand this prevents a distinctive charge rate for mobiles, generally causing mobile users to have to pay for airtime on incoming as well as outgoing calls. On the other hand, commonality of mobile and fixed number ranges may make it easier to have inter-service portability between fixed and mobile numbers.

European mobile number portability has a quite different technical implementation from geographic and non-geographic portability for three reasons. First, the GSM cellular switching system (see, for example Reference 3) is complex and it was important to contain the costs that might have resulted from extensive modifications. Second, since the GSM system has the inherent intelligence to manage subscriber mobility and roaming, there were opportunities to exploit this functionality for portability. Finally, mobile portability involves not only the routeing of calls, but also the correct handling of non-call transactions, for example text messages (the Short Messaging Service, SMS).

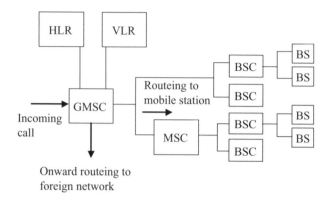

Figure 7.6   Mechanisms of normal incoming call delivery to a GSM network.

Figure 7.6 illustrates the normal handling by a GSM network of an incoming call from another network, be it mobile or fixed. Each mobile terminal or station has an assigned number known as its **Mobile Station Integrated Services Digital Number** (MSISDN). This is the number its user makes known for the purpose of receiving calls, and which callers dial. Originating and transit networks analyse early digits of the MSISDN to present the call to a **Gateway Mobile Switching Centre** (GMSC) in the called mobile network. The GSM network maintains a record of all its subscribers in a database known as the **Home Location Register** (HLR). One of its functions is to keep track of whether each mobile is currently on its home network, or whether it has roamed to another GSM network. If it is on the home network, the network will also have a record of the mobile in a database known as the **Visitor Location Register** (VLR), keeping track of whether the mobile is switched on, and if so the base station service area (these contain about ten transmitter cells) where it is situated. The incoming call is connected to the mobile via (if necessary) another **Mobile Switching Centre** (MSC), the **Base Station Controller** (BSC) and finally the serving **Base Station** (BS), the transmitter site. If the mobile is currently roaming, that is registered on a foreign network, then the HLR will contain a **Mobile Subscriber Roaming Number** (MSRN)[2], this being an international telephone number negotiated with the visited GSM network that secures connection to a port on a GMSC of the visited network and temporarily identifies the mobile. The GMSC makes an onward connection to this number in order to reach the called mobile. This is not, of course, a comprehensive description of GSM functionality, and the diagram shows for simplicity's sake many fewer BSCs and BSs than would exist in a real network.

Figure 7.7 shows the implementation of number portability for GSM networks. The essence of the solution is that while incoming calls are presented to the donor network as a result of digit decode of the mobile subscriber number, the query to the home location register is diverted to the equivalent register in the recipient network. This diversion is transparent to the gateway mobile switching centre, and is achieved by means of a Signalling Relay Function (SRF) provided at a signalling transfer

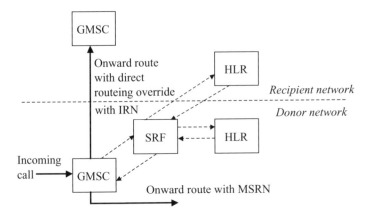

Figure 7.7   Mobile number portability in GSM networks.

point containing out-ported number data. This acts on SS7 **Mobile Application Part** (MAP) message between the GMSC and HLR by recognising the called subscriber as a ported number, and changes the addressed signalling destination to that of the recipient network HLR. Normally, the HLR query will return a mobile subscriber roaming number, to which the donor network GMSC onward routes the call. If the called mobile is roaming on a visited network, it is possible that the call may avoid all contact with the recipient network beyond the initial HLR query. This method in its pure form transits all calls via the donor network, that is, it behaves like an onward routeing solution, creating tromboned routeings should the call originate within the recipient network. It is possible to obviate this inefficiency by additional IN number screening beyond the scope of the GSM solution, and some (though not all) UK mobile operators use screening to detect and retain calls to their own in-ported numbers.

There are circumstances when it will be desirable or necessary to force the routeing of a call via the recipient network. The first is where the recipient network's customer requires a value added facility, maybe voicemail, that would be lost completely were the call not to pass through the network. The second is that with roaming calls, the donor operator may not wish to incur international call carriage costs on behalf of the competitor's customer, and should certainly be under no obligation so to do. Finally, the recipient network operator may not be content with reliance on the donor operator to provide billing records, nor with the donor network having the opportunity to track its customers' calling levels and patterns. To address these concerns, call routeing may be forced via the recipient operator's network at the initiative of either donor or recipient. This is known as Direct Routeing Override. If this takes place by the will of the donor operator, the signalling relay function returns an internally stored Intermediate Routeing Number (IRN) instead of the MSRN; this has the effect of routeing the call to the recipient operator's GMSC for onward handling there. The recipient operator can instigate direct routeing override by returning an IRN in place of the MSRN when the HLR is queried. Within the UK, IRNs are formed from the subscriber number (MSISDN) with a prefix indicating the operator, whilst the Netherlands, where 100 per cent screening in all networks is assumed, uses unmodified MSISDNs.

The European Telecommunications Standards Institute (ETSI) has prepared some standards for the implementation of mobile number portability with GSM [4, 5].

7.5   Regulatory strategy for number portability

7.5.1   Background

Number portability is viewed in many markets as a success. Geographic and non-geographic number portability have been introduced in a number of countries, with mobile number portability less widely available but growing. Some countries such as the UK now have over a million geographic ports in operation, with operators satisfied that it has stimulated competition. Customers, often unaware that portability may have been a contentious issue, expect it almost unconsciously. Incumbents have, of course, lost customers as a result of portability, but many see positive dimensions as well, first in that bilateral portability gives them a chance to win (or regain) customers from competitors, and second through the opportunity it gives them to play in the market for screening and transiting calls.

Nonetheless, portability has not been easily won, for four reasons. First, regulators have had to graft the requirement into licensing and rule frameworks that did not anticipate it. Second and understandably, incumbent operators have been disinclined to welcome portability, seeing it, at least at first, as an obligation to divert management time and invest capital mainly to facilitate the loss of customers to competitors. Accordingly, many have resisted it, being prepared to litigate both over the basic requirement and around the commercial arrangements for charging. Third, the development of technical solutions and operational procedures for portability has been well beyond the capabilities of regulators, who have needed to nurture co-operation between the network operators to generate jointly owned solutions. Finally, the commercial arrangements between operators and levels of charges have proved contentious and often acrimonious in many countries.

Once 'converted', incumbent operators have often taken leading roles in getting number portability working. To some extent this has been behaviour of enlightened self-interest, since they have been able to exert major influence on technical methods. They, perhaps, stood to lose the most from an externally imposed solution. For some, this has been a difficult balancing act. While half-hearted co-operation would certainly be perceived as anti-competitive obstruction, so also might zealous contributions seem to some an abuse of the dominant position. BT went beyond the requirements of their licence in offering out-porting of numbers a year before they could in-port them, even though the requirement called only for portability on a reciprocal basis. Ameritech, one of the US leaders in implementing local number portability, realised that proactive support of local number portability assisted their quest to be allowed to enter the long distance market.

7.5.2   *The terms of the portability requirement*

The simplest and most direct way by which regulators can mandate number portability is to make it an explicit requirement in regulatory rules or operators' licences, as in the UK. In most cases when first introduced, this is an amendment to existing rules and so faces all the normal hurdles of judicial challenge and of negotiation between regulator and operators. Other ways to introduce the portability requirement are:

- as an interconnection requirement (as when first proposed in Australia);
- as a numbering requirement;
- as a competition requirement (as in New Zealand), specified either positively, or negatively by classifying failure to provide it as anti-competitive behaviour;
- as a consumer requirement.

A choice of approach may have hidden consequences. Classification of portability as an interconnect service may, for example, imply the same cost recovery rules as agreed for mainstream interconnect, whilst a separate rule approach permits portability charges to be determined separately.

The form of the requirement to provide portability may be bilateral, universal, conditional, or subject to individual negotiation. New Zealand's light regulatory framework embodies the last of these, where an operator need not provide portability except by individual negotiation with another operator, or if a competitor successfully mounts a legal challenge under the Commerce Act on the grounds of unfair restriction of competition. The UK's initial requirement was bilateral, that is, an operator was obliged to provide portability to another operator requesting it, provided the other operator was able and willing to provide it reciprocally. This implied that the UK primarily regarded portability as a competition issue, leaving individual operators to decide when and whether they wished to seek portability with the incumbent, BT. More recently, however, the UK has had to change this when transposing European Directives into UK regulations. These express a consumer's right to demand portability. As a result, all UK operators must now be able to offer portability before contracting with a customer for service. The United States' regulator, the Federal Communications Commission (FCC), in the light of the 1996 Telecommunications Act listed local number portability as one of a number of conditions a local operating company had to fulfil before it could be allowed to enter the long distance market. Most regulators require operators to provide periodic returns to the regulator of number porting activity to help them manage the system effectively. It is unusual for there to be a requirement to accept in-ported numbers, since this is to an operator's advantage and so does not require regulatory intervention.

The portability requirement is normally separately applied in respect of different services, with most countries tackling portability in the following order of priority:

- geographic ('normal') telephone numbers;
- non-geographic services;
- mobile services;
- personal numbers, the so-called **Universal Personal Telephony** (UPT).

7.5.3   Choice of technical solutions

Regulators like to distance themselves as far as possible from technical solutions, partly because they lack the resources to develop them, partly because some are denied a technical remit in their constitutions, but mainly because regulation of network technology is very intrusive and they would rather leave market players to make their own decisions. Nonetheless, because portability solutions are by their nature co-operative, regulators cannot usually avoid some commitment to a solution, using wherever possible a co-regulation approach, which facilitates operator working parties to develop and take ownership of solutions. In most cases, this has been a successful approach. A danger that occasionally arises is when one of the players comes with a pre-commitment to a solution that has been adopted by it (or its equipment supplier) in another market. In the main, however, technical solutions are rarely major problems, it being the commercial terms that generate strong contention.

The choice of technical solution for geographic number portability has often been conditioned by desire for speed of realisation. Onward routeing solutions have lower capital cost and time-to-service at the penalty of higher running costs, whereas IN solutions are more heavyweight 'up-front' approaches that produce more efficient and possibly fairer long-term solutions. Some countries try to obtain the best of both worlds by combining an 'early' onward routeing solution with good intentions of later transition to an IN solution. It is all too easy for the IN upgrade to be lost, however, since the economic justification for the second change may be less persuasive than the original case for implementing portability. The simple call diversion solutions adopted at first in New Zealand, for example, are still in use. The Netherlands adopted an IN approach without prefix as its primary solution, giving a slower time to implement but a flexible environment allowing operators many choices between the costs of screening or paying more interconnect charges for less efficiently routed calls.

The United States has mandated the IN solution. Two factors lie behind this, a strict interpretation of a non-discrimination requirement, and a desire to implement number pooling in the same timeframe. Most countries' portability regulations insist that calls to ported numbers must receive the same quality of service as to non-ported numbers. This requirement for non-discrimination has been taken very seriously in the USA and led to an IN solution that screened 100 per cent of calls, since onward routeing and query-on-release solutions impose a slightly longer set-up time on calls to ported numbers than calls to non-ported numbers[3]. Other countries, such as the UK, acknowledge this difference but hold it to be insignificant and unlikely to influence a customer's choice of service provider or decision to port. Exhaustion pressure on the numbering system provided an adjacent factor in the US marketplace, where the IN screening technology for local number portability was also applicable to number pooling. Screening for number pooling allows them to spread a 10,000-line number block over two or more local operators' exchanges, so minimising granularity loss and thus conserving numbers.

The UK, after six years' experience of an onward routeing approach to geographic portability, is revisiting its technical solution [6]. This responds to two developments.

First, the basic success of number portability means that the number of ported numbers is no longer marginal, calling into question the wisdom of continuing with this technical solution. More urgently, however, the breakdown through bankruptcy of UK operators in December 1998 and in November 2001 focused attention on the unacceptable failure of portability in these circumstances. Customers with out-ported numbers were faced with a need to take a new number to continue service. There was no readily available solution for another operator to inherit the number blocks of the failed operator. In contrast, the USA was able to exploit its IN solution to secure rapid service recovery despite the total destruction of several exchanges by terrorist attack on September 11th, 2001.

7.5.4 Commercial arrangements

The operational costs of number portability for any operator fall into four classes:

- the set-up costs of technical and operational solutions for number portability, together with associated costs such as modification of customer care and billing systems;
- the per-occasion costs of processing each number port;
- the network running costs of screening calls;
- if applicable, the additional carriage costs of calls routed into the donor network and needing to be forwarded to the recipient exchange.

A number of general principles may be applied to the recovery of these costs. These may sometimes conflict, requiring balanced judgement on the part of the regulator. First, costs should fall on the agent that causes them. Second, costs should fall on parties who have an incentive to minimise them. Third, costs should where possible capture externalities. As an example, it would be better for costs to spread in proportion to operators' network sizes or market shares than, say, equally on large and small alike. Fourth and obviously, the arrangements must be practicable to apply. The solutions generally adopted around the world are as follows.

Operators bear their own costs of setting up the number portability capability. This is seen as a cost of participating 'membership' in the national competitive market. The cost captures to some extent the externality that it is likely to be proportionate to the operator's size. Onward routeing solutions may be preferable in this regard by not imposing a large up-front cost on all operators, although smaller operators can reduce the capital costs of IN screening solutions by sharing databases, or outsourcing screening to a transit operator. It is fair for operators to bear their own costs of modifying customer care and billing systems, since the burden of doing this may reflect their own care and investment (or lack thereof) in them over the years. It is probably not fair to insist that the entire burden of portability fall on the incumbent operator's shareholders or customers, as distributed benefits of portability will flow over the whole national telecommunications market. Regulators recognise this by allowing the portability infrastructure to be added to the network valuation base when working out long-run incremental costs for wholesale and interconnect rates.

The per-occasion cost of porting a number is normally borne by a direct payment from recipient to donor operator. Since the donor operator has no incentive to minimise, and market power to maximise this, it is usually price-regulated. Most countries allow some element of this cost to be transferred to the porting customer as a contribution to the benefit of porting. The UK regulator Oftel threatened intervention when one new entrant operator set its customer contribution at a discouragement level. This operator appeared to prefer to ignore portability, so frustrating consumer rights. Operators may waive this charge at their discretion for particular customers, although significant market power players have to apply it (or not) without discrimination.

The day-to-day costs of call screening are generally regarded like set-up costs as falling where they lie. These will usually be proportionate to the operator's size, and each operator has an incentive to be as efficient as possible.

The setting of per-call carriage charges by donor operators to recipient operators for onward routeing has proved extremely contentious. Should the donor operator be allowed to recover these costs in full, it may have little or no motivation to minimise them; conversely, should the donor operator be required to bear all these costs without recompense, then the recipient and other operators will have scant incentive to adopt efficient solutions themselves and may indeed welcome the opportunity to inflict costs on donor operators. A solution employed in the UK has been to set a regulated price with a time expiry, providing incentive for donor operators to migrate to more efficient solutions. Where IN solutions are employed, originating and transit operators may have choice over whether or not to use the donor network. In this case the donor operator's charges are avoidable and, being subject to competition, require less or no regulation.

7.6   Notes

1   This is because inter-exchange signalling in the UK uses full national numbers, the area code being inserted at exchanges when callers do not dial an area code. Hence, all user-dialled numbers begin with '0', or in special cases '1' for service codes and '9' for emergency calls.
2   This is one mode of operation. Some visited networks do not give static MSRNs, but assign a transient MSRN for each incoming call.
3   The difference is typically a few hundred milliseconds.

7.7   References

1   Confidential cost-benefit analysis for Oftel, 1993, quoted in 'Number portability: modifications to fixed operators' licences'. Oftel, 1997
2   'Economic evaluation of number portability in the UK mobile telephony market'. Oftel, 1997
3   BRYDON, A.: 'Global system for mobile communications – an introduction to the architecture', *British Telecommunications Engineers Journal*, 1995, **13**, (4), pp. 287–95

4 'Support of Mobile Number Portability: Service Description Stage 1'. ETSI, EN 301 715 (GSM 02.66) (ETSI, Sophia Antipolis)

5 'Support of Mobile Number Portability: Technical Realisation Stage 2'. ETSI, EN 301 716 (GSM 03.66) (ETSI, Sophia Antipolis)

6 'Consultation on proposals to change the framework for number portability'. Oftel, 2002

Chapter 8

Local loop unbundling and broadband services

8.1 Definition and overview

Local Loop Unbundling (LLU), shown in Figure 8.1, is the making available by a telecommunications operator of its physical customer access connections for use by other operators or service providers. These connections, typically thousands or millions of insulated copper pairs, provide the links between customers' premises and serving local exchange sites. Local exchanges provide the gateway to local and long distance services, and via indirect access to competitor network services. The UK incumbent operator, BT, has for example 43 million copper connections giving

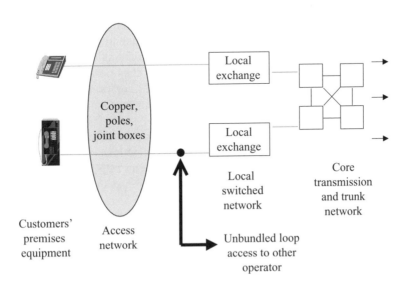

Figure 8.1   Typical incumbent network architecture showing the access network.

access to 98 per cent of the national population, while 28 million of these are in revenue-earning service. Such an access network has a replacement cost in excess of 20 billion pounds, representing a large and ubiquitous asset that is unlikely to be replicated and so gives its owner considerable market power.

Unbundling of various parts of the total service package offered by former monopoly operators has provided a means of introducing and sustaining competition, given that the scale economy and ubiquity of the former monopolist cannot readily be duplicated. Indirect access (see Section 5.3.4), for example, permits the unbundling of long distance service, allowing a customer to obtain this from a competing network while retaining the access and local services of the incumbent. Local loop unbundling of the copper pair service provides an opportunity for another operator or service provider to obtain all the facilities of the copper loop in by-pass of all the other incumbent services. The loop capabilities of particular and topical interest to other operators are its abilities to carry high frequency signals and so support broadband data transport rates. These would not be available to the indirect access operator, who depends on switched transport services with defined and inherently limited 64 kbit/s bandwidth.

Interest in loop unbundling developed during the 1990s as a result of the development of **Digital Subscriber Loop** (DSL) technologies, allowing high data rates over normal copper pairs that would have seemed impossible as recently as 1980. These offer prospects of switched video and home entertainment services, fast Internet access and broadband private circuits at prices within reach of the ordinary consumer. The United States' 1996 Telecommunications Act and also regulations in Europe and elsewhere have mandated loop unbundling, in the hope that this would stimulate both competition in the market for residential and small-business broadband services and the absolute growth of the market for these services. Loop-owning incumbent operators can use DSL technology themselves to provide broadband services, and many have opted for early market entry. However, it was felt that opening up the local loop to competition would result in earlier, more vigorous and more varied broadband market growth and so greater consumer satisfaction. Unbundling of the incumbent's local loop does not contradict the industry models of countries like the UK that had aimed for full facilities competition including access competition between operators, but simply activates an extra path for accelerating the coming of a mass-market information super-highway.

Local loop unbundling brings new technology into the telecommunications services industry and so raises a host of issues about the optimal way to grow this part of the market and encourage competition to that end. This calls for critical discretion on the part of regulators as they both develop new rules and apply old ones.

8.2 DSL technology and applications

8.2.1 DSL technologies

The Digital Subscriber Loop (DSL) family of technologies makes possible the transmission of data at rates as high as 50 Mbit/s over copper access pairs that had primarily

Table 8.1 *DSL data capabilities over different cable lengths.*

Data rate (Mbit/s)	Loop length (km)
1.2	5.5
2.0	4.8
4.0	4.0
6.0	3.6
20	1.4
25	1.0
40	0.3

been designed for analogue voice transmission at up to 3.4 kHz. The technology is extremely sophisticated, making use of the very latest in digital signal processing and very large-scale integrated circuit (VLSI) techniques. Excellent overviews of the technology may be found in References 1–3. The key technology milestones include:

- modulation techniques;
- adaptive transmission and reception techniques.

There are no hard and fast limits on the rates that can be achieved over a given distance, partly because of variations between flavours of DSL technology and different modulation techniques. It is not possible to tell in advance what data rates may be achieved over a given copper loop, because lines show a great deal of individual variability. Data from References 2 and 3 and shown in Table 8.1 illustrate what might be achieved in practice. Note that about 75 per cent of US local loops are within 5.5 km route distance of a local exchange, and about 80 per cent of UK local loops are under 5 km in length.

8.2.2 The DSL family

There are about a dozen DSL flavours on the market supporting different line configurations, applications and data rates. Most of the variations are specified to a fixed data rate, and the supplying operator simply predicts (or discovers!) whether for a given line the product does or does not work as the case may be. Although this technology can be rate-adaptive, that is it can determine its own maximum operating bandwidth on a given line, this capability has limited value to consumers. Most will prefer a fixed proposition, wanting to know in advance what they are buying and how it will perform. The DSL family consists of three main branches.

The **Asymmetric Digital Subscriber Loop** (ADSL) system provides a high downstream rate in the region of 1–6 Mbit/s from exchange to customer and a lower rate in the return direction of 16–640 kbit/s over one copper pair. This it can do in the presence of baseband telephony, or in other words the 'normal' use of the telephone line can continue even in the presence of the ADSL signals. These asymmetric data rates are suitable for switched video, home entertainment and Internet browsing applications, making ADSL the target product for the residential and home-worker market segments and the most widespread flavour of DSL.

The **High bit rate DSL** (HDSL) family provides symmetric data rates (the same in either direction) in the region of 2 Mbit/s, sometimes using two copper loops, one for each direction of transmission. HDSL is not normally compatible with ongoing use of the line for ordinary telephony, and so takes over the entire line. HDSL is a common choice for providing high bandwidth private circuits to medium and small businesses.

Very high bit rate DSL (VDSL) aims at very high symmetric or asymmetric bandwidths, of 13–52 Mbit/s downstream and 1.5–2.3 Mbit/s upstream, but only over short copper loops, usually less than 500 m for the higher rates. A major application is home video distribution, although business applications are also envisaged. The short length of reach prevents the use of VDSL over full loops from local exchanges, and it would more normally be used in conjunction with a **Fibre-To-The-Kerb** (FTTK) distribution system using copper only for the last drop to the customer's home. VDSL may be compatible with ongoing baseband telephony over the line.

Other members of the DSL family are listed in Table 8.2. Note that the claimed capabilities especially relating to distances are indicative rather than absolute. They are garnered from a variety of sources. No two lines are the same, so operation may both be possible outside a claimed range or impossible within it in individual cases. The line length from customer to exchange is rarely the radial, 'as the crow flies' distance. Suppliers improve their products over time, and may have made different assumptions about the noise characteristics of lines when predicting the likely reach.

8.2.3   DSL applications

DSL technology delivers to the customer premises higher bandwidths that can be used for a variety of applications.

- Video distribution, including games and education applications.
- High speed, 'always on' Internet access.
- High rate private circuits, for example T1 or E1 links, at a fraction of the traditional cost.
- Extension of an in-building local area network over a wider area to other premises occupied by the customer.
- Delivery of a total service package of telephony, data services and Internet connectivity over a single bearer.
- Bulk delivery of service to the manager of the wiring of a multi-tenanted building.

The first four of these are largely self-explanatory. The DSL line from the exchange terminates possibly at a filter to separate the telephony and high-frequency components and then on a DSL 'modem', or set-top box for TV based applications. Modems typically support USB ports for consumer and domestic applications, or Ethernet ports for business applications. The fifth of these is a new business opportunity, where a

Table 8.2   The DSL 'family'.

Main branch	Family member	Description
ADSL	ADSL	Asymmetric data over one pair: 1.5–6 Mbit/s downstream, up to 640 kbit/s upstream, ranging to 3.5–5.5 km depending on rate. Analogue telephony is simultaneously available on the line.
	ADSL Lite	Asymmetric data over one pair: 1.5 Mbit/s downstream, 384 kbit/s upstream. This was intended to avoid the need for a splitter at the customer end (though this has proved an optimistic assumption in practice) and so make self-installation ('plug and play' DSL) easier to achieve. Analogue telephony is simultaneously available on the line.
	UADSL	Universal ADSL – another name for ADSL Lite.
	RADSL	Rate-Adaptive DSL. This is like ADSL, except that it finds its best speed when initialising. RADSL has a symmetric mode running at 800 kbit/s in both directions.
	1-Meg modem	A proprietary variant of ADSL by Nortel.
	CDSL	'Consumer' DSL. A proprietary variant of ADSL by Rockwell.
	EZ-DSL	A proprietary variant of ADSL from Cisco.
HDSL	HDSL	High bit rate DSL. Symmetric data at 1.5 Mbit/s (T1) or 2 Mbit/s (E1) on two copper pairs at up to 3.5 km. Using an HDSL doubler, 7.3 km can be reached.
	HDSL2	An enhanced version of HDSL which uses one copper pair for 1.5 Mbit/s in both directions.
	SDSL	'Symmetric' DSL. Symmetric data at 768 kbit/s on one copper pair, or higher rates up to 2.3 Mbit/s over a range up to 4.1 km (1.7 km for highest rate).
	SHDSL	'Single line' HDSL. A new standard offering 192 kbit/s–2.3 Mbit/s symmetrical data over a single pair up to 4.7 km (2 km at the highest rate). Double these speeds are offered if two pairs are used.
	MDSL	'Moderate rate' DSL. Symmetric data at 768 kbit/s on one copper pair at up to 6.4 km.
	IDSL	'ISDN' DSL. Symmetric data at 160 kbit/s on one copper pair at up to 5.5 km.
	MVL	'Multiple virtual line'. A proprietary variant of HDSL from Paradyne, offering symmetric rates of 128–768 kbit/s in increments of 64 kbit/s up to 7.3 km. Alone among the HDSL family, this system supports simultaneous analogue telephony on the line.
VDSL	VDSL	Very high speed DSL. Asymmetric high speed data over one pair: 12–50 Mbit/s downstream, 1.5–2.3 Mbit/s upstream ranging up to 1.4 km but much less for the higher rates. Symmetric options are also possible.

service provider offers a total telecommunications service package of ordinary voice telephony, switched data services such as X25 or Frame Relay, Internet access and digital leased lines over the unbundled loop. This uses an **Integrated Access Device** (IAD), terminating the copper loop and incorporating the DSL modem. It possesses interfaces to ordinary telephones or the office PBX, the office local area network and the wireless local area network if any. Managers of multi-tenanted buildings may terminate bulk services from the local telecommunications operator and distribute facilities and capacity from the IAD to the various tenants. This prospect is probably more topical in the USA than in Europe, partly because of the greater prevalence of multi-tenanted buildings there. Another reason is that building owners in the United States often acquired the title to internal telecommunications wiring from the local operator at deregulation, gaining an opportunity that does not exist so much in Europe, where incumbents retained the internal wiring they owned prior to liberalisation.

8.2.4   *Alternative technologies for broadband*

DSL technology over copper loops provides a powerful platform for the delivery of broadband service to residential and small business customers using existing plant. It may be assumed that large business and corporate customers will be less interested, since many will have high capacity links and even fibre access already installed. Nonetheless, DSL is not the only means of broadband delivery to the home and small business. Some copper loops will be too long or otherwise of too poor quality to support DSL. Other technologies include:

- cable TV systems;
- fixed wireless systems;
- satellite delivery;
- fibre distribution, either **Fibre-To-The-Home** (FTTH) or **Fibre-To-The-Kerb** (FTTK) with a final metallic drop to the home.

8.3   System architectures

8.3.1   *Basic DSL service architecture*

Figure 8.2 shows the basic architecture of a DSL exchange connection. It illustrates the ADSL case where the line supports basic telephony (POTS[1]) in conjunction with ADSL. The broadband signals, drawn from an external broadband network, are modulated onto the DSL signal and injected to the copper line at a **DSL Access Multiplexer** (DSLAM). The connection between the ordinary telephone network and the copper line is made at a POTS/DSL splitter/combiner, often (as shown in the diagram) though not necessarily integral with the DSLAM. The copper loop leaves the exchange building via the **Main Distribution Frame** (MDF), and connects with the customer premises by way of the copper access network, probably traversing one or more cross-connection points on the way. At the customer premises, the line terminates at the **Network Termination Equipment** (NTE) and serves sockets

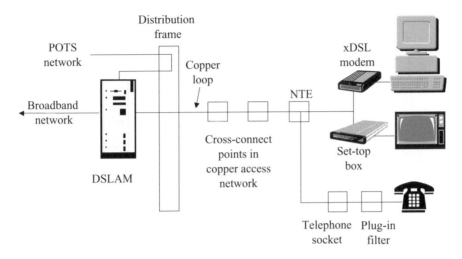

Figure 8.2   Schematic DSL system architecture.

giving connection to broadband and telephony customer premises equipment. Early architectures incorporated a splitter in the NTE, separating the DSL and baseband telephony parts of the signal from one another. Common practice nowadays is to use a self-install plug-in filter at each ordinary telephony port to prevent interference by and with the DSL signal. This saves the need for the operator to visit the customer site to change the NTE when installing DSL services.

The loop-owning operator may choose to provide a full end-to-end consumer service, including the broadband service, the copper loop and DSL bearer, and customer premises equipment. Competition may exist further back in the value-chain in the supply, aggregation and management of content in information and entertainment services. Further competition in broadband service provision over the local loop may take place in two ways. Under **wholesale DSL**, sometimes known as **bit-stream DSL** or by other marketing names, the incumbent provides the DSLAM and DSL facility but leases the consumer broadband data channel to another licensed operator or service provider. Under local loop unbundling, the loop-owning operator leases the physical copper pair to another operator or service provider who becomes responsible for the entire broadband service, including the DSLAM equipment.

8.3.2   Co-location and distant unbundling

Loop unbundling introduces a new element, **co-location,** into DSL system architectures. This is the logistical means by which a third-party operator obtains physical connection to the copper loop. The two basic methods, of **full co-location** and **distant location**, are shown in Figure 8.3. Under full co-location, the operator leasing the unbundled loop obtains space in the incumbent's exchange building, and in it positions a **Handover Distribution Frame** (HDF) and DSLAM equipment serving the

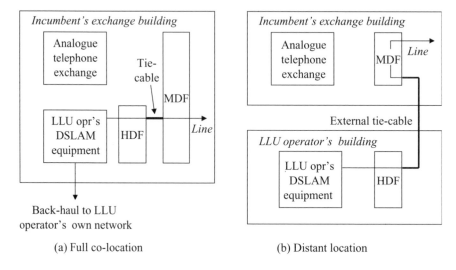

(a) Full co-location   (b) Distant location

Figure 8.3   Full co-location and distant location.

loop. The incumbent operator supplies a tie-cable linking its own main distribution frame with the HDF, and through this the other operator obtains access to the copper pair. In the case of distant location, the other operator houses its DSLAM equipment and HDF in a separate building, while the incumbent operator supplies an external tie-cable linking its MDF with the remote HDF. Although full co-location is inconvenient for most incumbents, regulators usually insist on its being made available, since distant location adds the length of the external tie-cable onto any loop used by the other operator. This would place it at a competitive disadvantage relative to the incumbent as it would then not be able to achieve the same rates or service-reach as the incumbent. Even with full co-location, the loop-leasing operator suffers unavoidably from the disadvantage that it has to provide backhaul capacity from the incumbent's exchange to its own network. This drawback may be avoided with distant location, but not necessarily since this depends on the choice of distant location, itself limited by the need to be quite close to the incumbent building.

Some unbundling operators use a nearby version of distant location by using equipment cabinets in the footway outside an incumbent exchange; this has in some situations provided a solution to problems in finding sufficient space within an incumbent building. Although the rapid conversion of national monopoly networks from analogue switching equipment to much more compact digital switching technology has resulted in a great deal of spare space in many buildings, some of them suffer from a shortage of usable space for co-location.

Incumbent operators may have four different approaches to providing co-location floor-space and accommodation for other operators.

- **Hostel co-location** provides a dedicated and walled area for one or more other operators to install equipment.

- **Co-mingling** takes place when other operators' equipment is installed alongside the incumbent's own equipment racks without physical separation.
- **In-curtilage** co-location applies when the incumbent supplies accommodation not quite within the building, but nearby and within the boundaries of its own site, for example in the car park.
- **Bespoke arrangements** may be agreed between the incumbent and other operator.

8.3.3   Full unbundling, sub-loop unbundling and line sharing

A copper loop may be unbundled in one of three ways: fully, partly, and on a shared basis. **Full unbundling** assigns the entire copper loop to the leasing operator, and is the form illustrated in Figure 8.3. It would be the natural mode of operation for the HDSL family of services, where the operation of DSL normally precludes any other use of the line. If chosen with an ADSL service, then either the LLU operator must take over responsibility for the baseband ordinary telephony on the line, or the customer must give it up. **Sub-loop unbundling** applies when some though not all of the copper loop from exchange to customer is leased to another party. This might be the typical mode of operation with the VDSL family, where the unbundled loop is taken from a cross-connection point within about 500 m of the customer, and fed there with optical back-haul by the leasing operator or service provider. Sub-loop unbundling is currently uncommon in most countries, though it is now a requirement of European regulations[2] in anticipation of growth of very high bandwidth services. **Line sharing** takes place when the incumbent retains use of the loop for its baseband, POTS service, but unbundles the higher frequency part of the spectrum for use by another DSL operator.

Line sharing poses a number of operational difficulties. The incumbent and unbundled loop operator need to co-operate in managing the line for two reasons. First, the customer, who must be presumed not to have diagnostic expertise, may report faults to the one operator when its clearance falls within the responsibility of the other. Faults may arise from interaction problems, even when the two operators' services work satisfactorily in isolation. Second, both have reasonable need for test access to the line, and activation of this entails the ability of either to break (or cause to be broken) the line at the splitter. During testing by either party, the service offered by the other also suffers interruption, so testing can only be done with contact and agreement between the two of them. This makes it desirable to extract as a separate unit the line splitter (often incorporated within the DSLAM), and a possible configuration for its provision is shown in Figure 8.4. One operator will own and control the splitter: agreement on which this is would hopefully be a minor issue.

Aside from the problems of joint management, shared line unbundling is a mixed benefit for the incumbent. On the one hand, line sharing is good for the incumbent by allowing it to retain the customer and the customer relationship. On the other hand, not to permit line sharing erects barriers to the entry of other LLU operators. These are twofold. If the customer wished to retain the incumbent POTS service, which he or she may want for the benefits of indirect access or other reasons, then without line sharing a second line would have to be bought. The unbundled loop operator or service

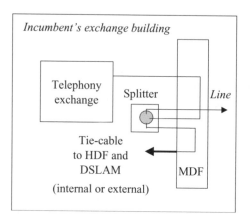

Figure 8.4　Configuration for unbundling with line sharing.

provider, who may have neither interest nor expertise in ordinary voice service, might find itself having to provide it in order to sign up customers who needed voice service and yet were unwilling to purchase another line. Most regulators, therefore, mandate incumbents to be willing to provide unbundling with line sharing where they continue to provide basic service. This is a European requirement.

8.4　Commercial aspects

DSL technology offers an attractive proposition to businesses in the form of data transmission at 2 Mbit/s or so over ordinary copper loops, which ought to allow them to make substantial savings on the cost of expensive leased private circuits, access to data networks or providing multiple exchange line capacity. Unfortunately for incumbents, who typically supply a majority of the private leased circuit market, a DSL service provides a substitute leading to loss of profitable and major leased line revenue streams. This will be a problem for the typical incumbent even if it provides the DSL service that threatens the leased line revenue. For the residential consumer, DSL is a relatively expensive technology where the equipment and set-up costs are a few hundred pounds per line. The service provider would most likely seek a service whose initial installation price was subsidised by ongoing subscription revenues. No 'killer application' has emerged, although certain opportunities such as switched video or Internet-based services seem worthwhile subject to the achievement of a critical sales volume or density of penetration.

　　The attitude of incumbents varies from country to country and has fallen into three broad categories. The first is obstruction, where the incumbent wishes to prevent or delay for as long as possible the arrival of LLU and DSL within its own or anyone else's product ranges. In these situations, the regulator has to act decisively. The second attitude is acceptance of the technology, though with a determination that it will dominate and control the market for DSL-based broadband products. This type of incumbent will resist to the maximum possible extent not the technology but competition and unbundling. The third type of attitude, possibly seen in Germany, is for the incumbent to recognise the new market, view competition as a means of growing the market to the common benefit, and co-operate in encouraging it.

A competitor seeking to enter the DSL broadband market by unbundling faces a tough business. Unbundled local loops are not cheap, and because incumbent retail line rentals are sometimes priced below cost even in liberalised markets, their 'fair' long run incremental cost (LRIC) prices as determined by regulators may be comparable with or even exceed retail rentals. This means that there is no great margin between wholesale and retail costs available to the unbundling operator, such as has been available in other areas, for example long distance call transport. The profitability of services based on loop unbundling depends, therefore, on the economics of the supported service. Competing operators face, besides the fixed costs of connecting with incumbent loops, the costs of back-haul to their networks. This adds to their per-line costs and places them at an unavoidable disadvantage when compared with incumbents whose exchanges are more than likely to coincide with network nodes or access points. These factors have led service providers in many countries to take a wholesale DSL product from the incumbent in preference to unbundling on their own behalf. A successful broadband services operator may then be one that blends wholesale products with full loop unbundling, carefully selecting its sites for full unbundling for a critical density of target customers within the right radius for the service in question.

8.5 Regulatory approaches

8.5.1 Requirement for unbundling

The basic regulatory approach to loop unbundling is to use licence conditions, typically insisting that players with significant market power provide some or all of:

- local loop unbundling;
- sub-loop unbundling;
- full co-location;
- line sharing.

The prices for loop unbundling have in most countries to be determined by regulators. An excellent and interesting review of pricing principles may be found in Reference 6. The prices include, besides the obvious ones of installation and annual rental for the metallic pair itself, the following items.

- Provision of tie-cables and ancillary services.
- Co-location accommodation and service charges.

The metallic loop installation and maintenance prices will differ for full and shared line unbundling, and may vary depending on the particular loop and the amount of work necessary to equip it for unbundling. Variant prices may apply according to whether the requested loop is to be newly provided or is an existing loop, and if so whether it already has the standard termination plug and socket. New provision is at its simplest if the order can be met from stock, but if not and various amounts of access network intervention and build are required, then higher prices may be levied. Few regulators will insist on unbundling provision should it entail major works of network build. 'Pair gain' systems, where two or more circuits are hosted using a multiplex technique on a single copper pair, are incompatible with local loop unbundling, and so a request for unbundling on such a loop will normally fall into the category of requiring new provision. Determined prices are normally based on the long-run incremental capital and current costs associated with providing that loop and any drop-wire, but not non-loop costs such as exchange line cards. Oftel required cost averaging between unbundling on exclusive copper loops and loops provided via pair-gain systems. Although this is a departure from individual cost-orientation, it was deemed appropriate to remove some of the risk the unbundling operator might encounter from cost uncertainty. The costs of co-location and external tie-cables where applicable include floor-space, building work, building maintenance, electricity supply, air conditioning, planning costs, doors, personnel access control, escorting costs, flooring and cabling, fuse-boards and smoke detection.

8.5.2   Strategic issues

The handling of loop unbundling by regulators requires a fine and delicate balancing of strategic issues, as well as discretion bearing in mind the local situation and the prevailing operator attitudes. The primary purposes of the regulator should be, as usual, first to ensure that consumers enjoy the best possible products at a reasonable price as soon as possible, and second as a means of securing the first, to encourage the optimum level of competition in the provision of those services. On the one hand, unbundling can stimulate the development of a soundly based, competitive broadband service market over the incumbent's access infrastructure. Unless, however, loop prices are very carefully weighed, it may encourage inefficient players into the market and deplete investment by the incumbent while simultaneously deepening its dominance.

European Regulations mandate metallic loop or sub-loop unbundling by operators with market power on a whole line or shared line basis, with non-discriminatory pricing [4]. National regulators have the power to ensure that the prices charged foster fair and sustainable competition. The United States 1996 Telecommunications Act required an incumbent local exchange carrier (ILEC) to unbundle at a regulated price any network element which, if not offered on an unbundled basis at the regulated price, would impair the competing local exchange carrier's (CLEC's) ability to compete [7]. This was at first interpreted to mean that any network element that can be unbundled must be, although this has been subject to plenty of litigation since then. The US position appears to be flawed for two reasons. First, it puts the competitor interest before a test of consumer welfare, and second it leaves undefined the meaning of 'impaired' and so does not provide a principle of limitation on the obligation. Ideally, it seems, any regulation requiring the unbundling of loops or any other network element should rely on the following tests.

- The unbundling is technically feasible.
- The network element is controlled by a company having significant market power in the relevant market.
- The requesting operator cannot reasonably duplicate the element in an alternative way.
- The company owning the element can exert market power by refusing to supply it.
- The company owning the element has refused to supply it at a cost-related price determined by a regulator.

Further questions that arise include the following.

- What is the right role for the incumbent in the broadband services market itself?
- How many entrants should be admitted to the market?
- Is it wise to be cautious in the introduction of innovative technology?

Incumbents have natural advantages in providing DSL services over their own copper loops. However, incumbent monopoly of DSL services is not an available option in many countries, for example under European and US rules. Nonetheless, an incumbent has opportunity to gain significant first mover advantage by bringing wholesale and retail DSL services quickly to market. If it did so, there would be positive and negative consequences to balance. A faster rollout of broadband services is a consumer benefit. However, there is a corresponding risk to the development of competition, were the incumbent to leverage its dominant position in the access arena into the broadband services market. The regulator must judge the extent to which it will sanction the incumbent's enjoyment of first mover advantage.

The regulator's preferred approach depends critically on the attitude of the incumbent. If the incumbent is an obstructor, then the regulator must do all it can to open up the market to more willing competitors. Requirements for incumbent unbundling should be backed by target timescales and penalties for failure to deliver. If the incumbent is a keen adopter, the regulator might well decide it was inappropriate or disproportionate regulation to slow this against the consumer interest while co-location arrangements were made for other operators to enter. It would, however, be appropriate then to regulate for transparent, non-discriminatory and probably controlled prices for the wholesale product.

DSL technology may have economies of scale and critical mass effects for profitability. It is therefore likely that fragmented competition would not exert significant pressure on an incumbent. A more promising market structure might be oligopolistic competition between a small number of suppliers of wholesale DSL service over the incumbent's loops, with a vibrant and competitive secondary market in service provision over the DSL paths. Should the regulator arrive at such a perception of the market, its choice would be over the extent to which it should, or could, engineer that structure into being.

In the UK, this issue surfaced over the allocation of co-location space in BT buildings, as this was expected to prove a scarce resource. The UK industry developed the 'Bow Wave' process (to be described below) as a scrupulously fair way of enabling as many competitors as possible to obtain a share in the market for loop unbundling. Oftel was called in to determine final details when agreement had not been achieved. Approximately 30 prospective unbundling operators dipped their toes in the market, though many later withdrew. Could it be argued that the attempt to facilitate the maximum number of market entrants had been an example of misregulation?

This is a fair question to ask, but it is worth considering whether any alternative courses of action existed. It is probably misconceived to argue that a regulator can or should operate affirmatively to limit the market players or winnow the number of entrants to a market. This would be very problematical, since to do so might entail a difficult selection process and a degree of intrusive market management. When markets are over-supplied or inefficiently fragmented, markets should, can and will correct this for themselves. That is a function and purpose of a market. The telecommunications services market was liberalised precisely to allow this to happen.

DSL technology is a new technology. Experts have real concerns about its technical viability at high penetrations. Should the market be allowed to gallop ahead, if necessary learning some lessons the hard way, or should conservative technical rules be applied, even if they imply delay and restraint? The technological question unearths the basic dichotomy of conservatism and caution versus experimentation and innovation, or in other words, 'How much risk is too much risk?' The UK developed an **Access Network Frequency Plan** (ANFP) through a co-regulatory working party, and in so doing was the first country to embody an ANFP in its unbundling regulation.

8.5.3   The access network frequency plan

Broadband carrier applications on copper pairs in multi-pair cables cause interference through cross-talk. This is a phenomenon where there is coupling of a signal on one pair into an adjacent circuit through the electric and magnetic fields produced whenever an alternating current flows. If the coupled signals are strong enough, they will corrupt and confuse the transmission. Unfortunately for broadband services, cross-talk increases with signal frequency, while failing insulation between pairs and imbalances caused by poor joints exacerbate the problem. The greater the number of pairs in a cable carrying such services, the more the effect and the greater the risk that both the service itself and other uses of the cable may be impaired, even to the point of becoming inoperable. The problem is a real one, with the potential to be very disruptive were it ignored until it manifested itself after considerable growth of broadband carrier services.

It is sadly not possible to completely determine safe operating limits either by theoretical modelling or by laboratory experimentation, since the real life characteristics of cables that may affect cross-talk and signal leakage (bad joints, ageing and insulation problems) will never be known exactly and may vary greatly from place to place. Some players in the debate took a conservative view and advised caution. Many of these were incumbents who could only too easily be accused of using this to obstruct

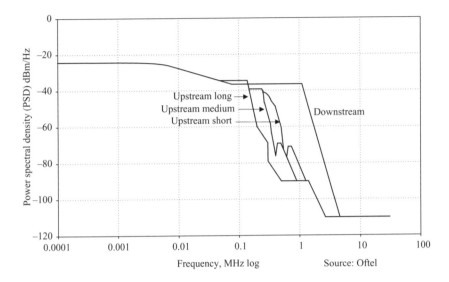

Figure 8.5   Indicative access network frequency plan.

DSL and unbundling, whether or not this was genuinely the case. Opposing arguments ran along the lines of, 'This technology is about to change the world, so you've got to take some risks,' or, 'We've not had any problems in California (or wherever).'

The difficulty of theoretically characterising this problem means, of course, that there is no right answer that can be defended to the satisfaction of all experts. The simplest approach, then, might be to locate the technologies currently in use and with the least disruptive interference characteristics, and specify these as preferred technologies. The rules would then prohibit the use of anything else. It would have been possible for the UK's DSL Task Group to generate an approved technology list, but they did not want to do that since it would have restricted future development. Instead, they devised a frequency plan in the form of a **Power Spectral Density Mask** (PSD mask) as shown indicatively in Figure 8.5, and described in References 8 and 9.

Each mask shows the maximum power that may be transmitted per hertz of bandwidth at each frequency. The three upstream direction masks labelled short, medium and long relate intuitively to line length though the categories are rigorously defined by a pair's insertion loss. The detailed nature of the masks simply reflects the spectral characteristics of the principal technologies already in use at the time, and has the effect of permitting any technology that sends less power than this at all frequencies. Opposing views on which technologies should lie within the adopted masks forced Oftel to determine the solution.

8.5.4   Co-location space allocation: a UK case study

In the early stages of local loop unbundling in the UK there were over 30 operators hoping to enter the market. The anticipated collective demand for space within BT's exchanges was therefore high and BT expressed concerns that it may not be able to meet this demand at a number of exchanges. Given this problem, the UK industry elected to develop a process for allocating space in situations where demand exceeded supply. This process became known as the 'Bow Wave' process[3]. However, industry was unable to agree on some of the final details of the Bow Wave process and subsequently requested that Oftel determine them. A similar concern about lack of space arose in Germany, but in contrast to the UK solution the German regulator (RegTP) ruled that Deutsche Telekom simply had to provide co-location somehow and would not be allowed to plead scarcity of space.

The UK's Bow Wave process did not prove a success, and is reported here not to advocate it as a solution for future use, but to illustrate the types of issues, problems and discussions that a national regulatory authority will from time to time encounter.

The Bow Wave procedure [10], which was very complex, ran in outline as follows. Each applicant could request space at up to 1,500 of any of BT's sites in priority order. This was a straight ranking with no nuancing, for example there was no way of expressing the notion that, say, 1–100 were equally important while 101 was very much less important. The Electoral Reform Society, a UK body noted for the conduct of elections of various kinds, handled the initial orders, and using a transferable vote system obtained a list of the 500 most desired sites. Of these, BT would target to provide co-location as fast as possible at 360 of these. This was the maximum number their resources could handle at the start, and the residual 140 were held as reserves lest any of the 360 prove infeasible for immediate development for any reason.

The applicants would get what they had requested at the chosen sites, excepting where there was over-demand. In that case, the successful applicants would be decided in descending order of the priority they had placed on that site, for example if a site had room only for one applicant, then an applicant placing it at priority 8 would take precedence over another that had assigned priority 15.

The voting procedure gave very limited opportunity for an applicant to quantify the requirement, and depended on an assumption that a basic unit of three equipment racks[4] was a worthwhile allocation for any applicant. Some players took the view that while this allocation measure was appropriate for business applications such as leased circuits and broadband data network access, it was insufficient for operators offering the 'triple-play' consumer services of video-on-demand or interactive television, Internet access and basic telephony. These felt they needed something like six racks as a minimum. The system at first allowed double bids (for six racks) at a site, but contained the possibility of the granting only of a useless (to that applicant) three racks. The six-rack issue led to much dissension. There was no incentive not to apply for six, was one complaint. 'Could one six-rack allocation be reserved at each site?' was another thought. The problem with this, however, was that it would have to be reserved for something definite, for example an applicant with a consumer business case. In that case, would the applicant be bound to that business case if it turned out to be unsound? Might Oftel find itself having to judge the worth of a business case to confirm the applicant's self-classification? If so, that would have represented a degree of intrusion that the UK regulator had tried to avoid. It was primarily over this issue that Oftel was called in to make a determination. When a consultants' investigation found that a three-rack allocation was reasonable for all applications, this issue was laid aside.

A potential flaw in the Bow Wave process, which had been foreseen by Oftel, was that it gave applicants little certainty of receiving enough space to justify their investment plans, or of receiving it in a sensibly complementary set of locations. Oftel had hoped that post-allocation trading of spaces would allay these concerns. However, this failed to satisfy and eventually large numbers of applicants simply withdrew their interest in the market. Frustrating though this procedure must have been, it is hard to believe that it was their only reason for market exit. As we have seen, the business of an unbundled loop operator is a challenging one. It is unlikely that so many would have left the market so quickly had they been really convinced there was money for the making.

The Bow Wave process ceased to be relevant after the market withdrawal, and one imagines that most of the people involved with it were pleased to leave it behind. Applications for co-location space with BT are nowadays proceeding on a normal first-come-first-served basis, and there has not so far been an issue of exhaustion nor contention for space.

8.6 Notes

1 POTS = Plain Ordinary (or Old) Telephone Service.
2 The European requirement for loop unbundling was first promulgated in 2000 by a Regulation [4], but not embodied in a Directive until 2002 [5].
3 The name 'Bow Wave' originates from the initial surge of applications expected on the opening day.
4 A 'rack' is a space of normal height (2.4 m) measuring 600 mm by 600 mm with a standard allowance of surrounding floor space for heat dissipation.

8.7 References

1 YOUNG, G., FOSTER, K. T., and COOK, J. W.: 'Broadband multimedia delivery over copper', *IEE Electronics and Communication Engineering Journal*, 1996, **8**, (1), pp. 25–36. (This is a shortened version of Reference 2)
2 YOUNG, G., FOSTER, K. T., and COOK, J. W.: 'Broadband multimedia delivery over copper', *BT Technology Journal*, 1995, **13**, (4), pp. 78–96
3 There is a collection of articles on Broadband Access Copper Technologies in *IEEE Communications Magazine*, 1999, **37**, (5)
4 Regulation (EC) No 2887/2000 of the European Parliament and of the Council of 18 December 2000 on unbundled access to the local loop (The European Parliament, Brussels, 2000)
5 Directive 2002/19/EC of the European Parliament and of the Council of 7 March 2002 on access to, and interconnection of, electronic communications networks and associated facilities (The Access Directive) (The European Parliament, Brussels, 2002)

6 'Access to bandwidth: indicative primers and pricing principles'. Oftel, May 2000

7 United States Telecommunications Act 1996, Section 252 (d)(2) (US Congress, 1996)

8 'Access to Bandwidth: Proposed solution for the Access Network Frequency Plan (ANFP) for BT's Metallic Access Network'. Oftel, June 2000

9 'Access to Bandwidth: Determination on the Access Network Frequency Plan (ANFP) for BT's Metallic Access Network'. Oftel, October 2000

10 'Statement and Determination on local loop unbundling 'Bow Wave Process'. Oftel, November 2000

Chapter 9

New products and services

9.1 Regulation in a changing world

9.1.1 Change and stability

A telecommunications services regulator has to regulate a rapidly changing market. Some of this flux is a result of the regulation itself, but not all. Were the advancement of competition the only change, then a regulator's task might eventually terminate after the transition from monopoly to stable competition. In reality, the market is exposed to technological evolution, product innovation and commercial realignments. The regulator must examine each of these, and react by continuing, withdrawing, amending or extending the intervention currently being practised. The aim in all cases will be to ensure that consumers enjoy the best products at the most advantageous prices, using competition as the primary medium of achieving these goals. The regulator must, therefore, act against actual or potential exercise of market power to the consumers' detriment.

Technological advances may create new markets, or significantly change the economics of an existing market. The following and similar developments call for regulators to look at market trends and determine the most appropriate form of regulation, if any.

- Digital Subscriber Loop (DSL) technologies, described in Chapter 8, open a new market for residential and small business broadband service.
- Mobile network operators, reaping economies of scale, achieved rapid growth by lowering prices. This transformed their business from a premium, professional product to a mass market.
- Mobile networks are beginning to compete with and substitute for fixed line service in certain market segments, for example teenagers and second lines.
- As intelligent network services became increasingly popular, a worthwhile secondary market appeared for non-network-operating players to offer value-added services.
- Internet access has silently revolutionised traffic patterns on public fixed-line voice networks.

9.1.2   Basis for regulatory intervention

The regulator must identify and analyse new markets created by the service, identifying any players who will possess market power. It must examine its impact on the degree of competition in existing markets. Consultations are normally necessary after initial analysis of the issues, and for novel services these may necessarily be extensive. Regulation concerning new services and developments may take one of the forms listed below.

- Imposition of an obligation or prohibition, usually on dominant players but possibly on others.
- Technical prescriptions, for example about interfaces.
- Imposition of price controls, usually on dominant players.

The legal mechanisms for intervention include:

- existing rules, and determinations and consents thereto;
- rule amendments;
- action under Competition Law;
- action under European Directives (in EU member states);
- issue of guidelines.

9.2   Dial-up Internet access

The use of the public switched telephone network (PSTN) for accessing the Internet using voice band data modems has risen from negligible proportions in 1990. In 1997, BT reported that its Internet traffic, then growing at 10–15 per cent per annum, accounted for 10 per cent of its 248 million daily local call minutes [1]. By mid-2001, a third of all public telephone network traffic in the UK was Internet traffic [2]. Much of this uses 'unmetered' access and exploits number translation services that had in the mid 1990s been the destinations of a marginal proportion of calls.

There are various ways in which customers obtain access to Internet Service Providers (ISPs) via the PSTN. The simplest form of access is via 'normal', geographic telephone numbers. If using this method, a national (or international) ISP would typically have an array of access numbers in various towns and cities to bring most customers within local call reach. Spurred by the growth of revenue-share number translation services, UK ISPs found they could more efficiently address their customers' access with 0845-type number translation services, for which callers paid a local call charge.

Countries with a tradition of unmetered local calls, the USA being a notable example, faced a tidal wave of incremental call minutes at zero revenue. Internet calls last a long time (various sources quote average durations of up to an hour) as compared with three or four minutes for ordinary voice calls. Pacific Bell forecast network collapse within 18 months of October 1996 [3]. Most European countries had metered local calls and did not face this problem initially. However, and probably as a result, their Internet usage did not grow at anything like US rates.

Competition was aggressive in the UK ISP market. Freeserve was in 1998 the first to offer a subscription-free service. Callers still paid for their telephone calls, the revenue-share portion of which generated Freeserve's principal income. In June 2000, BT introduced a retail unmetered Internet access package under the name **SurfTime** for its own and other participating ISPs. Customers paid a flat-rate subscription though they paid no call charges for access as this was via a Freephone number.[1]

Rival ISPs who did not participate in BT's retail product found it difficult to compete with SurfTime, since the only wholesale access products offered to them were metered products. Some of them did offer an unmetered consumer product based on an averaged subscription and a metered wholesale access product from BT. This exposed them to potential margin squeeze and so financial risk. Consequently, a number of them asked Oftel to intervene and compel BT to offer an unmetered wholesale access product.

Oftel analysed the market for dial-up Internet access [2], and reported that dial-up access was likely to remain the dominant mechanism of access to the Internet for residential consumers and small business for the foreseeable future. (Larger businesses typically use more expensive broadband data networks or leased line services.) The market was analysed into one retail market for Internet access and two derived wholesale markets, for call origination and call termination as shown in Figure 9.1. The analysis found that while the retail Internet access and wholesale call termination markets were effectively competitive, the wholesale call origination market was not competitive. BT dominated it, by virtue of the number of access lines it controlled: it had a market share of 81 per cent (by lines) or 79 per cent (by residential call minutes). They accordingly determined that BT should supply a wholesale **Flat Rate Internet Access Call Origination** (FRIACO) product. Because FRIACO is a flat-rate call origination product, its price to the purchaser is not in units of pence per minute, but of pounds per annum for capacity, per 64 kbit/s or 2 Mbit/s port. This was to be revenue-neutral for BT, in that it would raise neither more nor less revenue in total than would have been raised under metered supply. Prices were to be controlled under Oftel's Network Charge Control Regime, which in its 2001 proposals was evolving

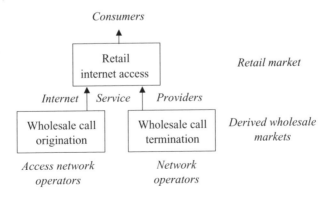

Figure 9.1   Market structure for consumer dial-up Internet access.

from separately determined long-run incremental costs (LRIC) to a five-year RPI-X price cap [4].

Because flat-rate access was expected greatly to increase call volumes, BT, the other market players and the regulator needed to work together on a satisfactory technical solution. The ideal technical solution is to detect the flat rate calls at the earliest point in the switching chain, the local exchange, and extract them straight to the terminating operator and thence the ISP. The most efficient way, and a possible long-term solution, would be to convert these to Internet Protocol (IP) at the local exchange. Modem banks and access servers at local exchanges would pass the data via an IP data network to and from the ISP's computers as shown as option (a) in Figure 9.2. Another solution, shown as option (b) in Figure 9.2, carries the Internet calls as *voice* calls between the local exchange and the modem banks at the ISP's premises. This solution uses Intelligent Network (IN) screening at the local exchange to detect calls to Internet access numbers, and routes them to terminating operators over direct interconnecting voice links. Both of options (a) and (b) in Figure 9.2 were not instantly applicable solutions, as both would have required terminating operators to extend their networks to have direct links with all of BT's local exchanges[2]. For an incumbent network operator such as BT, it was a necessary solution, since its trunk switching layer could not carry the expected call volumes without substantial investment.

The problem preventing wholesale adoption of local exchange handover of Internet calls, solution (b) in Figure 9.2, was that most other operators interconnected with BT's network at the transit, or DMSU layer. They wanted BT, therefore, to switch calls via its trunk units onto the interconnect links from there into their own networks. If forced to interconnect at local exchange level, they would have faced a great deal of new network build to obtain direct access to all BT's local exchanges. BT has about 750 of them, although there are fewer physical sites than this, and 340 are co-located with transit exchanges. Accordingly, when BT at first offered only option (b), a local exchange FRIACO product, this was held to be obstructive of competition. Oftel determined that BT must also offer a **Single-Tandem FRIACO** (ST-FRIACO) product [5], shown as option (c) in Figure 9.2.

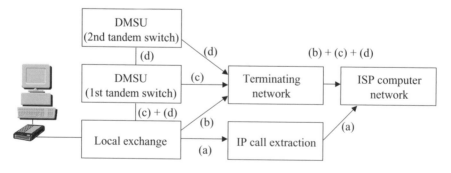

Figure 9.2   Handover options for Internet access call origination.

Consultants, called in to analyse the impact of FRIACO on BT's trunk network, examined BT's view that its trunk network could not bear the extra traffic. They evaluated the costs and investments at stake for network upgrading [6], confirming that the problem was real. Oftel's determination had thus to reflect both feasibility and fair treatment of BT in the event of its having to make network investment. A double-tandem FRIACO product, option (d) in Figure 9.2, was neither offered nor required. To insist on this would have been inappropriate regulation, as it would have allowed a regional or local ISP to obtain a national footprint on the strength of investments made almost entirely at BT's risk.

The technical solution consisted of three phases.

- Phase 1, a short-term solution, placed a capacity limit on BT's obligation to provide ST-FRIACO. Other operators had to make a corresponding commitment to migrate to local exchange interconnect where practicable.
- Phase 2, a medium-term solution, pre-supposed augmentation of the transit layer of BT's network.
- Phase 3, a long-term solution whose detailed outworking was deferred for further co-regulatory study, was for local exchange grooming of Internet traffic into an IP network. (At the time of writing[3], this phase is not now being progressed, since phases 1 and 2 appear to be providing a satisfactory solution.)

Phase 1 placed a total limit on the number of ST-FRIACO ports BT would provide, in line with existing network capacity. This solution was accompanied by a commitment on any operator taking ST-FRIACO to rearrange some of its FRIACO traffic (and, if it so chose, other types of traffic) to local exchange interconnect. The first and easier rearrangement was for the other operators to accept local exchange interconnect with the local exchanges that were physically co-located with tandem switches. Even this was not, however, free of cost to them, as it divided former single interconnect traffic routes into separated routes. These smaller routes would have lower utilisation[4] and require extra exchange and signalling ports. The second level of rearrangement was to use Interconnect Extension Circuits (IECs) from tandem locations to reach local exchanges not co-located with a tandem switch. This incurred the costs of the IECs in addition to the above. The operators had some discretion over which routes to rearrange and in which order of priority, but BT also had some power to require rearrangement. Both knew that adding traffic 'headroom' at a tandem switch by offloading to local exchange interconnection, would add to the limited ST-FRIACO capacity.

Phase 2 entailed build by BT of extra tandem layer capacity. This took the form of an overlay 'Internet' network to collect FRIACO calls from all the DLEs and distribute them to interconnect points at the existing tandem switch locations. This solution would relax the limitation on ST-FRIACO capacity, since the overlay network could grow as required. There was considerable discussion of the fairest way to recover the capital cost of this new network capacity. BT's problem was the considerable risk that the new switches might become stranded (prematurely obsolescent) after the implementation of the ultimately preferable IP network solution of phase 3. It being considered inappropriate to ask BT to bear this risk in its entirety, some equitable sharing arrangement was required. It was difficult to estimate the proportion of the risk that BT should bear, as no one really knew BT's spare network capacity precisely, nor the spectrum of alternative uses that stranded switch capacity might subsequently find. The full complexity of the arguments may be found in Reference 6.

The preferred solution, not finally determined at the time of writing[5], appeared to separate the network interconnection charges for FRIACO service into two parts. The first was the normal per port per annum charge, but reduced to exclude the capital elements that would normally be present. The second was an upfront charge that spread the entire capital cost of new capacity over orders for that new capacity and in proportion to the orders. BT itself would under this scenario bear investment risk in proportion to its actual use of the new capacity. This would be borne not in BT's network business but (under non-discriminatory network charging) by its retail Internet access business in equality with competitors. It appeared that a penalty charging mechanism might be necessary for 'late' ordering, to forestall the incentive for operators to under-forecast new capacity requirements. This would cause quality of service problems for all.

9.3   Mobile virtual network operators

A **Virtual Network Operator** (VNO) is one that markets a publicly available telecommunications service, but who pays another operator to provide some of or the entire necessary network infrastructure. A VNO could differentiate its service from the retail service of the underlying network provider by branding, customer service and billing, or possibly by bundling it with other telecommunications or non-telecommunications products. This has nothing to do with the **Virtual Private Network** (VPN) services offered by some network operators. These are virtual *networks*, while we are here discussing the subject of virtual *operators*. VNO interest in fixed network services appears so far to have been scant. Nonetheless, Oftel has proposed in its recent review of retail price controls on BT, that BT should offer a wholesale line rental product to support VNO provision using its network [7]. This proposal, to which BT has agreed, is expected to support competition in the access market where BT has dominance. There has been considerably more recent interest and commercial activity in virtual network operation in the mobile sector. Regulators have, therefore, to address what regulation if any is appropriate for **Mobile Virtual Network Operators** (MVNOs).

The definition of an MVNO is not at all clear. There is a range of options for the balance of facilities an MVNO would provide for itself and those it would take in unbundled form from the Mobile Network Operator (MNO). These are illustrated in Figure 9.3. The subscriber management function holds and manages the databases used for verifying subscriber identities, service details and access rights. All MVNOs would provide at minimum handsets and **Subscriber Identity Module** (SIM) cards. Having no spectrum allocation, they would rely on the MNO for the base station and air segment infrastructure. Such an MVNO would have a separate identity as a GSM operator (it would possess a Mobile Network Code, MNC), and could negotiate

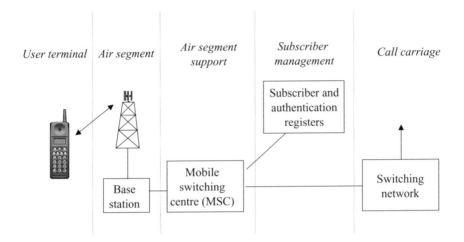

Figure 9.3   Virtual mobile network operator unbundling options.

roaming agreements with other networks at home and around the world. The MNO would recognise the MVNO's subscribers and connect them to the MVNO network from the earliest economic point in the MNO network. At maximum, the MVNO might buy the call carriage service and subscriber management functions as well as radio infrastructure from the MNO. However, if the MNO provided all this, it is unclear how the MVNO would then differ from an ordinary retail service provider and so whether the term MVNO would be sensibly applicable. The call carriage and subscriber management elements shown in Figure 9.3 are not a hierarchy and either could be taken or not taken. The MVNO might ask the MNO to switch some of its calls, but not all of them.

The prospective consumer benefits of MVNO supply are:

- greater competition in the retail market, hence more efficiency and lower prices;
- service innovation, especially in the form of Intelligent Network (IN) and information services;
- composite service offerings, notably combining fixed and mobile services from a single supplier.

The first of these advantages is at its maximum when MVNOs have freedom to change their underlying MNO supplier. They might even have concurrent parallel MVNO agreements with two or more of them, selecting the most efficient MNO supplier of capacity on a call-by-call or longer-term basis. The advantages to a mobile network operator are less clear. It does not have relationship with the MVNO's customers, and cedes the profit margin of those parts of the value chain that the MVNO provides. It would, presumably, view MVNO operation as beneficial only to the extent that it brought network utilisation that it would not otherwise have obtained.

Prospective MVNOs in the UK formed a view that the MNOs would be mostly unlikely to supply unbundled service without regulatory obligation so to do[6].

Examining the issue in 1999, Oftel [8] found insufficient grounds for intervening to compel mobile operators to supply unbundled service to MVNOs. The mobile market (in the UK) was classified as prospectively competitive. While accepting that there were potential benefits (as above) from the extra competition that MVNO operation might introduce, Oftel believed that these consumer benefits might come about under existing market conditions. There was a concern that mandating the provision of unbundled service by MNOs for MVNOs could lead to reduced infrastructure investment. This was because MNOs sometimes invest in rural capacity that is not in itself economic, in order to differentiate their service by coverage. This advantage might be eroded or lost completely if MVNOs could switch seamlessly between networks. Concluding that MVNO operation on the one hand entailed investment risk and on the other was not a necessary condition for new and innovative services, Oftel decided that intervention to compel MVNO support would be disproportionate regulation.

9.4   The Intelligent Network (IN)

The **Intelligent Network** (IN) is a term that loosely describes an architecture for added value services in public switched telecommunications networks using associated computers. The software controlling the telephony switch detects a trigger point, often though not necessarily at the digit decode of a dialled number. This initiates a signalling interchange with a service programme running on a different computer. The service programme supplies information to guide the connections that the switch makes. The services typically supported by the IN principle are:

- number translation services (including Freephone);
- dynamic routeing support for call centres;
- virtual private networks;
- alternative billing services (including calling cards);
- personal numbering services;
- unified messaging services;
- call screening services.

It may seem odd at first sight, given software control of digital switches, that such an architecture should be necessary. However, it enables added value services, including speculative and lightly-used services, to be rolled out rapidly over whole networks without frequent and expensive modification of the ultra-reliable control software in each exchange. It may also be the most economic solution to provide some highly used services in this manner. Two families of standards for the signalling interface between exchange and service computer evolved during the 1980s and 1990s. These are a North American ANSI, and an ITU-T, set of standards. Although different vendors' implementations are theoretically interoperable within the family, most use proprietary extensions that restrict practical interoperability.

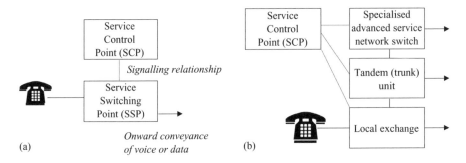

Figure 9.4   Simplified Intelligent Network (IN) architecture.

The IN architecture implements a functional separation between the network layer that routes calls, and a service layer for service logic. In the liberalised environment, it raises the possibility of competition within the service layer. Companies that do not own network infrastructure could participate, therefore, in an advanced services market.

Figure 9.4 shows, in highly simplified form, the IN technical architecture and its implementation in a typical network. Part (a) of Figure 9.4 shows the main elements of the IN. These are the **Service Switching Point** (SSP), or switch that triggers as required to access the advanced service logic, and the **Service Control Point** (SCP) containing the service logic itself. A network need not equip every single switch as an SSP, but has three deployment options illustrated in part (b) of Figure 9.4. In many networks, these options follow in an evolutionary sequence as IN services become more widely used. The first step is usually to provide IN service access at dedicated exchanges, typically 'service nodes' that are specialised for the purpose. When a call requires an advanced IN service, the local exchanges (and trunk tandems where necessary) bring the call to one of these nodes. These secure the onward connection in accordance with the advanced service logic. As a next step, the network operator typically rolls out SSP functionality to its trunk level exchanges, this stage being current in many European incumbent and competitor networks. The final stage is to provide SSP functionality in some or all of the local exchanges, a stage more characteristic of United States networks. The first two stages have lower costs relative to the final stage in terms of the number of switches that must support and maintain IN functionality. However, they incur higher costs through the potential for inefficient call carriage via SSP exchanges regardless of the ultimate destination. Network operators make a calculated trade-off of these costs when forming their network strategies.

A third-party service supplier who wishes to provide an advanced service but does not own a network has, at least conceptually, four available tactics. The first, option (a) in Figure 9.5, is to own an SCP that will be accessed from other networks. This method is rarely used, however, because it is fraught with technical problems. The Intelligent Network signalling standards between SSP and SCP were designed for

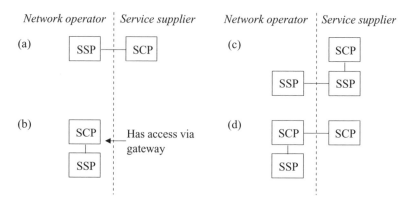

Figure 9.5   Third-party service provision in the IN.

an era when networks were monopolies, and one company had control of both ends of the interface. The SSP-SCP interface could threaten network integrity if deployed between equipment whose owners do not have a trusted relationship. The second tactic, option (b), is for the network operator to have tight control of the SCP but to allow third parties sufficient access to mount services upon it. This is technically though not commercially comfortable. The network operator and service supplier may compete in the same advanced services market. The network operator gains a great deal of commercial intelligence about the other service supplier's services and usage profiles. Some network operators exploit a variation of this approach, allowing their customers access to their SCP for real-time control of call routeings, or customisation of advanced service logic.

The commonest form of third-party service supply is shown as option (c) in Figure 9.5. The service supplier possesses its own switch (SSP), which gives access to its IN service. This approximates to the specialised switch strategy shown in Figure 9.4, where in this case the third party supplier owns the specialised switch. This raises few technical problems or interoperability issues, but does require the service supplier to be an authorised network operator or reseller having interconnection with the network operator. The service supplier suffers a competitive disadvantage relative to the network operator, since calls using its service must bear the overhead of call transport to its switch. Typical services today have enough added value to bear this as a small impact on margins.

Option (d) provides an open interface between operators at the service level. This is technically advanced; pregnant with development possibilities it is more evident in laboratories than commercial reality. It allows the SCP of the network operator to acquire an image of the service logic from the service supplier's SCP. The network operator's SCP might act as a simple agent or gateway for messages to and from the distant SCP. Alternatively, it could download a temporary copy of the service, like a Java application on the World Wide Web. A current implementation of this principle is the **Customised Applications for Mobile network Enhanced Logic** (CAMEL) capability in mobile networks. This allows a roamed customer to obtain use of a special service from his or her home network while connected to a visited GSM network. A recent initiative to update the service control interface in telecommunications networks is that of the PARLAY group [9]. This is an open consortium of operators, equipment makers, software vendors, service providers and applications developers.

The principal current activity for regulators concerning IN services is not large, and is to ensure that network operators with dominance do not leverage that dominance into adjacent service markets that could be competitive. These powers have been used to ensure, for example, that fixed and mobile operators with market power permit the third-party supply of voice-messaging and unified messaging services.

9.5   Next generation networks and services

9.5.1   *Voice over Internet Protocol (VoIP)*

It is possible to carry voice telephony calls over the **Internet Protocol** (IP). **Voice over Internet Protocol** (VoIP) thus forms an alternative transport technology for the traditional telephony service, which is based on the provision of continuous, both way 64 kbit/s channels between the connected parties. IP is a network layer protocol for the transfer of variable length data packets, and its emergence as a near-universal standard has made possible the worldwide interconnection of computers, now known as **The Internet**. IP packets may transmit standard-quality telephony, as long as the link bandwidth and spare packet capacity are sufficient to support the constant, bi-directional 64 kbit/s data flow. Useful collections of technical papers may be found in References 10 and 11.

VoIP may or may not save bandwidth, and this effect is a complex balance of various factors. Every IP packet carries at least 20 bytes of address data and other overheads, which is of course wasteful. Three techniques serve to lessen the bandwidth load, though all have quality penalties. Lower quality products may, however, find willing customers when deeply discounted. Voice compression down to 8 kbit/s and less provides usable though not standard quality, and may add considerable processing delay. Silence suppression, that is, not sending packets during speech silences, saves a lot of bandwidth but is vulnerable to the clipping of the beginnings of words. Accumulation of speech samples to form bigger packet payloads improves the payload to overhead ratio at the cost of delay. It takes, for example, three milliseconds at 64 kbit/s to assemble 24 eight-bit samples, yielding a 55 per cent ratio of payload data to total data sent. This amount of delay would not in itself threaten the usability of a telephone connection, but it adds to other delays that may be present. Were the total to approach about 20 ms, call quality would be unacceptable.

Common applications of VoIP and other packet-based techniques such as **Voice over Asynchronous Transfer Multiplex** (Voice over ATM) include:

- computer to computer voice calls over an Internet connection;
- voice transmission over company data networks, both on the local office network and over the wide-area corporate network;

- voice transmission over leased circuits, possibly for resale;
- operators' bulk voice transmission over national and international backbone links.

It appeared that at the end of 2000, about 6 per cent of worldwide voice calls used a VoIP link at some point. About 20 per cent of these were computer-to-computer calls, and 25 per cent of calls were made with a calling card. The most enthusiastic adopters of VoIP were new entrant operators keen to diversify into voice at minimum cost, while the least willing were the former incumbent operators. Many of these were conscious of quality concerns.

Regulators normally aim for technological neutrality, neither imposing nor favouring particular implementations. The use of VoIP, therefore, may not raise regulatory issues when used as an underlying transmission technology in an operator's backbone or corporate internal network. Since computer-to-computer communication is generally not offered as a public service, a VoIP call within this context might not receive regulatory attention, except in two matters. Computer communication may be subject to general content regulation, for example that it may not contain obscene or offensive matter. Some countries might view computer-to-computer voice calls as an illegal by-pass of a monopoly telephone network and evasion of tariffs. However, were an operator to launch a publicly available alternative switched telecommunications network using VoIP, then technological neutrality implies that it should be subject to exactly the same regulation as any other PSTN service. This might include obligations to provide, for example, emergency service, directory enquiries and operator assistance. The operator might be obliged to negotiate interconnection terms when asked, to ensure basic service interoperability and to conform to essential interface requirements. Oftel's current VoIP guidelines [12] say that these obligations would only apply to a VoIP service if it were considered to be a 'publicly available telephone service' as defined by these characteristics.

- It is sold as a substitute to a publicly available telephone service.
- It appears to the end user as a publicly available telephone service.
- It provides the only method of accessing the traditional circuit switched telephone network.

9.5.2   Next generation network technology

9.5.2.1   Overview

It appears that telecommunications network technology is on the threshold of a major evolutionary change. Comparable in magnitude with the transition from analogue to digital transmission in the 1970s and 1980s, this changeover is from the provision of constant bandwidth connections, known as circuit switching, to packet-based transmission and routeing. Whereas today's networks provide telephone calls with a reserved bi-directional bandwidth of 64 kbit/s between the two points engaged on a call, tomorrow's networks will, if this change eventuates, instead switch packets like the Internet.

The underlying driver for this change is the exponential growth in data and computer communication, which accounts for ever-growing proportions of the total bandwidth in world telecommunications today. Figures are out of date the moment written, though industry estimates talk of voice comprising less than 10 per cent of total global traffic by 2004, and less than 1 per cent by 2010. These figures are by volume, however, and voice remains commercially predominant for many operators in terms of revenue and profit. Before, data were the minority application and so modems adapted data for transmission over a voice infrastructure. As the table turns, voice will be the minority application, to be adapted by packet technology to travel as the guest on a world data network. New networks following these principles are known loosely as **Next Generation Networks** (NGNs).

Technical architectures for NGNs have probably now stabilised at the outline level. The reasons for change to NGNs, which are explored below, appear clear and compelling, at least at a theoretical level.

- Network generalisation: multi-purpose networks using packet transmission technology.
- Network flexibility: low functionality transport networks with peripheral intelligence and service layers.
- Low cost networks.

The arrival of next generation networks presents a great challenge to the industry and its regulators. The current generation of networks are a legacy from the age of monopoly. Their interoperability and service quality originate from a painstaking process of standards development that took a long time, and was allowed to take a long time. Next generation networks are growing up in a very different, liberalised and competitive world. Interoperability and service quality have not yet been secured, while standards processes display some fragmentation.

The investment boom culture of the 1990s has contributed to a great deal of hype and misinformation. Among the many jockeying for key positions may be found:

- 'guru' prophets claiming to have a 'glimpse of the future';
- equipment manufacturers claiming mastery of the future, no matter what that future might be;
- salespersons, dealmakers and entrepreneurs imaging themselves as Internet pioneers and visionaries;
- start-up companies engaged in investor confidence play, with eyes focused first on share option, public offering and buy-out hopes before sustainable business.

Industry culture is arguably less supportive of standards development than before. Boom culture has been neither comfortable nor rewarding for the individual who simply wants to do a good job well. These are the talented people needed to make fine judgements, pick their way through detail and make systems and people work together. In a culture of image and presentation, hype and misinformation compound uncertainty. The phenomena described in 1985 by Amos Joel, computer controlled digital switching system pioneer, are present realities.

> [The state of this technology today] is being influenced a great deal not by solid technical achievements, but by heralding and black box descriptions that portend great technological accomplishments. The trade press, which should know better, is party to the curtain of

mysticism, clichés, and cacology around which they shroud the true technology of new products or groups of products. It is unfortunate that this industry, whose past has so benefited from the free exchange of technical information, has been reduced to this state in the name of competition. [13]

We can hope that new generation networks will be successfully launched and implemented, and on a serious scale. Maybe they will. Maybe they will crystallise over the next 20 years, emerging more slowly than suggested in extravagant propaganda. Maybe the technology will contain features that re-invent circuit switching in all but name. Maybe regulators will need a heavier hand in the regulation of interfaces for interoperability.

9.5.2.2   Packet-mode networks for generality

A packet network is a flexible, general purpose network platform for delivering multiple services. Packet transmission is capable of carrying any communication, and so generalises the applicability of core networks. This continues an established trend. Analogue networks were typically application-specific, needing separate transmission infrastructure for each service, such as 3.4 kHz voice telephony, telex, broadcast quality voice, TV distribution and so forth. Digital transmission, introduced in the 1960s, offered a universal technology that could carry everything over a single digital core network. However, the network generalisation brought about by digital operation was not complete, since the networks and switches were optimised for constant bit-rate services such as leased circuits, telephony and video. Computer communication has very different dynamic characteristics, such as burstiness and asymmetry. Packet networks can, in principle, support all these templates with only a small overhead for packet addressing.

9.5.2.3   The low-function, 'pure' transport network

The ideal network for a rapidly evolving, competitive multi-service communications industry is a network that does little more than pure transport of data. More ideally, it might be a passive network where data are injected at the input point and emerges at its destination without need for an explicit switching function. Separate servers connected at the network periphery then provide the applications logic that provides useful services to people. These can come, go and evolve quickly or slowly according to need. This is a very different approach to traditional telephony networks, whose core switches contain embedded functionality related to voice switching. This feature simultaneously optimises today's networks for the voice service, but makes them less efficient or adaptable for other types of service. A vision for new architecture is explored in more depth in Reference 14.

Figure 9.6 elaborates these ideas. Pure passive networks exist, but are unfortunately not scaleable. One example is a **Passive Optical Network** (PON), where all input is relayed to all destinations[7]. To achieve communication, an input and an output agree on corresponding wavelengths or timeslots, and communication happens. Wireless transmission provides a similar environment. Some form of control plane is necessary to find the transmission path and notify it to the parties that wish to

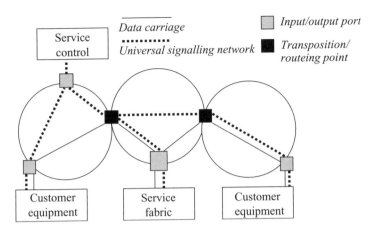

Figure 9.6   Conceptual framework for next generation network.

communicate. A scaleable and modified version of a passive network architecture is illustrated in Figure 9.6. Passive routeing domains, shown as circles in the figure, interconnect at transposition (or routeing) points where some switching functionality is needed to link channels between domains. A control server, connected like any other terminal at the network periphery, controls routeing. It communicates with end-user terminals and routeing points using a **Universal Signalling Network**. This is itself hosted in allocated bandwidth within the network. If a network service needs to add value to the data flowing between end-users, this is achieved by routeing the data flow through a service fabric, again at the network periphery. Such a service might be data adaptation (for example, standards conversion), information supply, or telephony call control. The network architecture is almost completely virtual, as the data flows in the network have no meaning and structure except those given them by the service logics. Such a network is capable of rapid functional evolution with few bounds.

9.5.2.4   Internet implementation of the transport network

The currently envisaged implementation of the 'passive' network architecture for the Next Generation Network is based on Internet technology. Packets find their way through routers to the destination, guided by the IP address in the packet header. The routeing process may be viewed as passive to the extent that no con-figuration of routers is necessary in respect of any one packet or communication session between two end-points. The Universal Signalling Network is formed by exchange of IP packets between terminals, fabrics and servers. A worldwide IP net-work architecture based on a hierarchy of network levels appears to be emerging. Figure 9.7 illustrates the principle, showing two layers. A backbone upper layer con-tains high-capacity links and routers; these are interconnected for resilience but are not necessarily fully interconnected. A lower layer, replicated regionally, mirrors

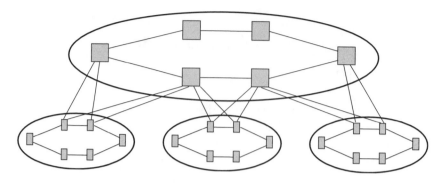

Figure 9.7   Next generation network hierarchy.

the basic upper layer structure and has links between at least two routeing nodes and at least two nodes of the upper layer[8]. The emerging global network may have three layers: a global core layer, a national routeing layer and a local distribution and access layer. The global layer and national layers may employ optical switching technology for very high bandwidth and so require a sub-structure of optical and electrical routeing planes. Of course, layering is not rigid and the architecture permits flexibility.

There is a selection of papers on carrier-scale IP networks in Reference 15.

9.5.2.5   The low-cost network

The next generation network is a lower-cost network than traditional circuit-switched networks, partly because of its intrinsically lower cost per bandwidth unit, and partly because of the whole-life cost advantages of a unified, evolvable platform. The intrinsic cost of switching bits per second through high capacity Internet routers is about one twentieth of that when using traditional telephony switches. This is not, however, a true cost comparison for voice *service*, since call processing service logic and the assurance of adequate quality of service must be added to basic packet transport. Taking this into account, next generation network technology may provide voice service at around half the cost of circuit switching technology.

9.5.2.6   Next generation and current networks

The interconnection and interoperability of current and nascent next generation technology are essential as long as both exist side by side. Physical interconnection is most likely to use 2 Mbit/s trunks as shown in Figure 9.8. The circuit-switched exchange perceives the next generation network as though it were another digital switch, and communicates with it using the inter-exchange Common Channel Signalling System No 7 (SS7), thus passing call control details. The next generation network requires two elements to support a voice service. The first is the **Media Gateway**, which transports the voice data to and from the network, performing the conversion from

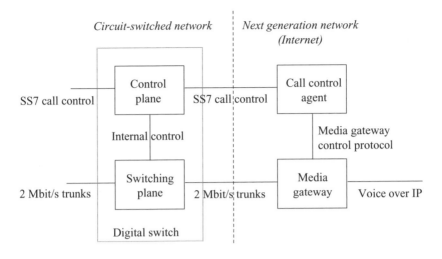

Figure 9.8 Interworking of next generation and current networks.

the 64 kbit/s or other voice standard to packets. The second is the Call Control Agent. This provides essential call processing functions such as ringing bells, giving service tones, alerting and clearing, decoding dialled digits, choosing the destination address, and billing. It communicates with the circuit-switched network using standard inter-exchange signalling (possibly adapted to flow over an IP network layer), and with the media gateway using a **Media Gateway Control Protocol** (MGCP).

9.5.2.7 Next generation networks in practice

The next generation network is not yet a serviceable replacement for current voice networks. This is because its flexible and relatively unstructured nature does not guarantee sufficient quality of service for standard-grade voice calls. An apt analogy is the comparison between railways and roads. A voice call is like a train, in this case of speech samples. The circuit switched network reserves and clears a path for the entire train for the required journey. Excepting for disasters like breakage of the path, the train is sure to arrive, and to arrive on time. The packet network requires each speech sample (or sample group) to find its own way like a car through the network of road junctions, that is the routers. Each packet's journey time is unpredictable, depending on what else is on the road at the time, and there may be queuing at the junctions. The cars may all have different delays, they may arrive in a different order to their starting, and some may not arrive at all. Scaleable and interoperable solutions for bandwidth reservation in packet networks, necessary to resolve the quality-of-service problem, remain to be deployed. When operators build next generation networks, and for reasons stated above many feel a pressure to say they have done so, their precise implementation may be much more limited in scope, and more realistic, than a full conversion of their network to Internet Protocol.

9.5.3   Regulation and next generation networks

Regulation of next generation networks has tended in many countries to be light or non-existent for a trio of reasons.

- Use of next generation technology within carrier networks may not surface at interconnection interfaces, and escape regulation if seen only as an internal transport mechanism.
- Internet services are often regarded as competitive and so subject only to light regulation if any.
- Thoughts of intervention may be tempered by a fear of stifling innovation.

Nonetheless, the prospect of wider deployment of next generation network technology contains dangers that regulators need to address. These revolve around the quality of service issue, and are:

- a risk of market dominance;
- compromised interoperability between networks.

It is possible that next generation network operators may side-step network standards issues by interconnecting using the well-known 64 kbit/s circuit switched bearers, with common channel signalling for call control. This will not be sufficient, however, to safeguard interoperability on account of the time delays that build up in the conversion of voice for packet transmission. Cumulative delays, which will be worse the more conversions there are, could make voice quality unacceptable.

An unregulated Internet is in danger of generating and perpetuating a measure of monopolistic supply. Unlike traditional circuit-switching with its International Telecommunications Union (ITU) standards, next generation and Internet technologies have yet developed little established understanding of the service quality that a network can expect (or demand) from an interconnecting partner. It follows that only an operator with end-to-end control of a telecommunications link can offer a user-to-user quality-of-service guarantee. This gives a large advantage to the global network operators, who may obtain joint dominance and exercise extensive market power. There may already be signs of this, in the occasional unwillingness of large Internet backbone operators to interconnect with smaller operators. In Internet jargon, this is known as 'peering'. The former often feel that peering gives subsidy to the latter. It may be appropriate for peering to become mandatory for players with market power, on commercial terms that follow the usual principles.

The standardisation process needs to be encouraged, as the fast-to-market environment sometimes leaves it poorly co-ordinated and fragmentary. This follows partly from the 'boom' culture, where both individuals and companies may find it more rewarding to promote glamorous topics, even with limited effectiveness, in preference to hard and detailed standardisation work. Proprietary variation of standards is a tempting option for a supplier anxious to secure a first mover profile. It is also a tool of market power, when a supplier with control over an interface attempts to lock customers into its platforms and products. As next generation Internet technology becomes more pervasive and so more of an essential facility, then it is reasonable to require some freezing of standards for inter-carrier interconnection. Network management aspects are easily ignored in the rush for 'innovation', yet standardisation here is vital for realistic operation and operational network support.

As voice networks converge with data networks, a common regulatory approach is appropriate. It will not serve consumer interests to treat next generation implementations separately simply because the name 'Internet' enters their description. Because the Internet is a global network, then no single national regulatory authority will be able to cope unless there is global co-operation. National regulators need, nonetheless, to keep an open eye for conflicts and preferably develop co-regulatory frameworks to oversee national network development.

9.6 Other issues

9.6.1 Radio spectrum

Radio spectrum is a classic scarce common property resource, needing national, regional and international regulatory oversight for its effective exploitation. The licensing and control of spectrum has existed for many years, and the reader is referred to other sources for technical and historical detail, for example Reference 16. Established agencies for spectrum regulation, often government departments, have continued this work during and after telecommunications liberalisation. As a result, many national telecommunications regulatory authorities do not handle spectrum issues directly[9].

The historic and largely universal method of allocating spectrum resource has been by central planning. Spectrum licences issued to named operators are usually highly specific to the point of stating the transmitting sites, stipulating the technical standards and prescribing the use to be made of the spectrum. Licences that are limited in number may be awarded first-come-first-served, or by analytic evaluation of applications. The latter process, known as a 'beauty contest', cannot guarantee economically efficient outcomes that maximise producer and consumer well being. This is not primarily because of corruption or ineptitude (though these can happen), but because of the informational burden of obtaining the best solution. The current regime makes it difficult to trade or reassign the uses of an allocated spectrum. It gives little incentive for holders of a spectrum to release what they do not need or could be more effectively used by someone else. It is, however, an effective mechanism for guaranteeing that spectrum will be available for defence and the emergency services' requirements.

Market-based mechanisms of spectrum allocation might produce more efficient allocations. A detailed discussion may be found in Reference 17. This implies maximum access to spectrum with competitive market pricing mechanisms. The highest value uses might then migrate to the highest value spectrum. An allocation mechanism should aim to maximise *economic* efficiency measures and hence consumer well being. This would not necessarily optimise *technical* efficiency parameters such as users per megahertz. A mass-market rollout of a cheap and spectrally inefficient technology might, for example, produce greater total well being than an expensive spectrum-conserving technology that confined a product or service to a premium market. A spectrum market might possess the following features.

- A spectrum property right that is technically neutral and unencumbered with usage prescriptions.
- Technical restrictions on spectrum use only when necessary to prevent harm to public safety, public health and other spectrum users (or where international treaty obligations exist).
- Auctioning as the mechanism for selecting holders for limited spectrum rights.
- Rights for spectrum holders to trade, sell or sub-lease their spectrum rights.
- Reservation of spectrum for public good purposes (e.g. defence, emergency services) not excluded, but subject to opportunity cost evaluation in the light of market prices.
- Regulatory provisions to detect and offset market failure. For example, players designated with significant market power might require specific authorisation to trade spectrum.

The UK was the first country to hold an auction for spectrum for Third Generation Mobile Services. Held in April 2000, this raised £22.5 billion for five licences. This and auctions elsewhere raised a total 110 billion euro within Europe [18]. Sums of this magnitude have resulted in serious debt problems for European operators. Questioning commercial realism, some regard this outcome as raising doubts over the viability of auctioning. Nonetheless, the parties, who in many countries had visibility of their competitors' bidding, entered their bids freely. If the high prices were in part due to artificial restriction on the amount of spectrum or number of licences for 3G use, then this adds rather than weakens support for free market approaches.

9.6.2   Location services

Telecommunications networks possess information about the location of their customers. This includes both static information such as billing address and fixed line address, and also dynamic information such as the calling line identity (CLI) and in mobile networks the current base station or transmitting sector area. Triangulation technologies, notably the **Enhanced Observed Time Difference** (E-OTD) method using enhanced mobile handsets, allows the current position of an operating mobile handset to be determined to about 50 m accuracy. E-OTD does not rely on satellite positioning systems and has the potential to be a mass-market technology.

Location information, where a user has no desire to keep it secret, makes possible a number of beneficial services. Geographic location information has commercial value to information service providers, as it allows them to customise the information sent to the caller's location. The making available of this to external applications is normally a matter for commercial negotiation between network operators and information providers. It would become a subject of regulation if a dominant operator were leveraging its position in a way that restricted competition in information services to the detriment of the consumer.

The US regulator, the Federal Communications Commission (FCC), has mandated the making available of location information by networks to the Public Service Answer Points (PSAPs) of the emergency services in two phases. Phase 1 covered calling line identity, or base station identity for mobile networks. Phase 2, targeted for 1st October 2001 and full implementation by 31st December 2005, required 67 per cent of callers to be identified to within 50 m and 95 per cent to within 150 m. This target was a piece of adventurous regulation and a challenge to operators, since it was of uncertain feasibility. A number of carriers have filed for waiver of the 1st October requirements. European regulation is less prescriptive, requiring location information to be delivered to emergency services 'to the extent technically feasible' [19].

9.6.3 Roaming and call terminating charges

All network operators possess a monopoly in the market for call termination, since no other operator can complete a call to one of its customers. This presents an opportunity for monopoly pricing. Apparent examples are the mobile operators who charge much more to other operators to terminate calls than to their own subscribers, even though the resource cost is substantially the same or less. Mobile roaming charges, notoriously expensive within Europe and reported to yield a 90 per cent margin for the operators [20], are similarly exploitative as the customer has little opportunity to avoid them (except by taking a new subscription with an operator in the visited country). Both have been subject to regulatory investigation over the years, and Oftel, for example, has demonstrated that pricing is above the competitive level. Customers of mobile networks have very limited incentive to move away from operators that charge highly for call termination, since they do not pay the charges in question.

A typical mobile operator would typically defend high call termination rates with the following arguments.

- 'Regulation of call termination prices is inappropriate because the profits subsidise lower subscription and outgoing call prices, so benefiting consumers overall'.
- 'Call termination is subject to competitive pressure, because callers can call fixed line numbers instead, while owners of closed user groups can change their operator'.

Oftel's latest view, revealed in Reference 21, is that the market for call termination is unlikely to be competitive in the foreseeable future, and that price regulation is therefore appropriate and justified. The offsetting of excess profits in one market against lower prices in another is not economically efficient, nor is it fair to consumers. Because the (UK) mobile market is not fully competitive, it cannot be assumed that the excess termination profits are fully competed away in the subscription and outgoing call markets. Finally, while substitution pressures do exist in the call termination market, these are judged insufficient to bring termination charges to a competitive level. The preferred regulation was by price cap (of Retail Price Index −12 per cent) until 2006[10].

9.6.4   Premium rate services

Premium rate services are information services that cost more, often very much more, than a normal telephone call. They often include **revenue share**, where the service provider receives a share of the cost of a call, this providing its revenue stream. Some services are separately billed, however: accessed by a 'normal' telephone number, the service provider recovers revenue under a separate billing relationship with the caller by invoice or credit card. Premium rate services include simple information (e.g. weather, share prices, tourist information), entertainment, phone-in competitions, advice lines and live conversation ('chatline') services. The majority of premium rate offerings give good service to their customers. Nonetheless, this sector has a twilight zone where dishonest operators abuse consumers. Regulation of premium rate services is thus appropriate, addressing the following issues.

- Assisting consumers to control their expenditure on premium rate calls.
- Control of consumer abuse by dishonest operators.
- Control of access to these services, for example by children.
- Control of the promotion of, and the content offered by, these services.

A suitable regulatory framework for premium rate services may have two principal elements. The first is an industry self-regulatory code-of-practice for each service type. In the UK, the Independent Committee for the Supervision of Standards of Telephone Information Services (ICSTIS) provides this. However, this needs the backing of a secondary regulatory power to order disconnection of non-conforming services. Experience shows that this second power is essential, since some operators have been slow or unwilling to disconnect, even in blatant cases. Oftel also has a power to recognise or not recognise a relevant code-of-practice as suitable. Concluding in 1992 that the code-of-practice for chatlines gave insufficient consumer protection, Oftel used its powers to withdraw recognition of the code-of-practice, effectively prohibiting this type of service in the UK. Expenditure and access controls include:

- rights for individual subscribers to opt into or out of access to all or various types of service;
- Personal Identification Number (PIN) access control;
- setting of individual credit limits for premium rate services;
- notification when expenditure reaches pre-set thresholds;
- clearly distinctive numbering ranges for premium rate services.

9.6.5   *Telemarketing and automatic call-making devices*

Many organisations use **Automatic Calling Equipment** (ACE). This is call-generating equipment that allows them to call a large number of people at very low cost. Usually employed for selling and marketing purposes, this results in a large number of unwanted telephone calls. Many of these calls do not result in live speech. Some operators employ **predictive dialling**, where a computer pre-dials calls and assigns an operator only when a call is answered. As there is a finite probability of there being no operator free at the time of answer, this may result in someone being called by silence. Some operators have used ACE to 'look for' fax machines, disconnecting when a person answers. Such calls may be mistaken for malicious calls or telephone stalking. There is a valid issue whether regulation is appropriate to protect consumers from these calls. Against the benefit of suppressing irritating and nuisance calls must be set the development of innovative uses of the telephone network.

Regulation varies from country to country, and may differ from actual practice in the market because of enforcement difficulties. Regulations may separately control different call categories, for example speech calls and calls that do not result in live speech. Three of the possible regulatory approaches are as follows.

- Customers must opt, in writing, before receiving calls.
- Consumer consent may be required but not necessarily in writing.
- The field is open except where customers specifically opt out.

Written opt-in is rare for telemarketing calls, because it would in practice prevent the market from developing. Non-written consent may rely on implicit consent, calling for judgement, and there may be enforcement difficulties. British regulation, enshrined in operator licences and class licences for ACE operators, was until recently unrestricted for speech calls originated by ACE but required written consent for calls which did not mature into live speech. This latter category included the playing of recordings. These conditions were enforced in 1998 with a final order against an operator who had been making large numbers of auto-dialled calls to identify fax numbers. A recently proposed amendment [22] retains the requirement for consent for calls not maturing in live speech, but removes the need for it to be in writing.

Further rights may arise under Privacy Regulations, these being adjacent provisions and not part of telecommunications regulation as such. In the UK, individuals have the right to opt out of receiving marketing and sales calls, whether made by people or machines and whether directed at telephones or fax terminals. To take advantage of this facility, people have to register their telephone or fax numbers with a Preference Service.

9.7   Notes

1   The particular Freephone numbers used in the UK were in the 0844 range.
2   We refer here to BT's approximately 750 local exchange nodal 'processor sites', and not to the 6,000 historic local exchange sites that are mostly served by remote concentrator units. Further details of network architecture may be found in Chapter 5, Section 5.2.2.
3   July 2002.
4   This follows from the 'Erlang' effect, where the more the traffic carried on a route, the less the proportionate overprovision of capacity necessary to secure a given grade of service. Section 4.2.2 contains a graphical presentation and more

detailed discussion of the 'Erlang' effect. The minimum provision quantum, of a 2 Mbit/s link supporting a block of 30 circuits, additionally and seriously cuts into efficiency when smaller routes are required.

5   July 2002.

6   An MVNO operator may reach a commercially negotiated agreement with an MNO without regulatory intervention, and there is such an arrangement in the UK.

7   This type of network architecture requires, of course, that thought be given to coding and modulation schemes that will maintain privacy and confidentiality of communication.

8   Beware of diagramming conventions! Many Internet architecture diagrams show the backbone layer at the bottom of the page. This is the reverse convention to pictures of traditional networks, where the highest (or backbone) layer is at the top. This latter convention applies in the figure.

9   The UK's impending reorganisation of telecommunications and media regulation will bring Oftel's work under a common organisational umbrella, Ofcom, with spectrum regulation.

10 The mobile operators rejected this proposal and Oftel referred the matter to the Competition Commission. Its conclusion in January 2003 imposed a steeper degree of price reduction.

9.8   References

1   'Phone use grows as more prefer surfing to talking', *Financial Times*, 13th March 1997, p. 22

2   'Oftel's 2000/01 effective competition review of dial-up Internet access', 30th July 2001, paragraph 2.2

3   'California surfing could flood telephone networks', *Financial Times*, 25th October 1996, p. 1

4   'Proposals for network charge and retail price controls from 2001'. Oftel, February 2001

5   'Draft Direction on future interconnection arrangements for dial-up Internet in the United Kingdom'. Oftel, November 2000

6   'Consultation on future interconnection arrangements for dial-up Internet in the United Kingdom'. Oftel, November 2000

7   'Protecting consumers by promoting competition: Oftel's conclusions'. Oftel, June 2002

8   'Oftel Statement on Mobile Virtual Network Operators'. Oftel, November 1999

9   *www.parlay.org*

10 Collected papers on Internet Telephony, *IEEE Communications Magazine*, 2000, **38** (4)

11 There is a collection of articles on Voice over IP in *BT Technology Journal*, 2001, **19** (2)

12 'Frequently asked questions on the regulation of Voice over Internet Protocol services'. Oftel, April 2002

13 JOEL, A. E. Jr.: Guest editorial in *IEEE Journal on Selected Areas in Communications*, 1985, **SAC-3** (4)

14 BUCKLEY, J. F.: 'Switching and service delivery in futuristic networks', *BT Technology Journal*, 1993, **1** (4), pp. 64–72

15 There is a collection of articles on Carrier-scale IP networks in *BT Technology Journal*, 2000, **18** (3)

16 WITHERS, D. J.: 'Radio Spectrum Management' (IEE Books, London, 1999, 2nd edn.)

17 'Oftel's response to the Independent Spectrum Review of Radio Spectrum Management'. Oftel, September 2001

18 Quoted in *Financial Times*, 13th June 2002

19 Directive 2002/22/EC of the European Parliament and of the Council of 7 March 2002 on universal service and users' rights relating to electronic communications networks and services (Universal Service Directive), Article 26. The European Parliament, Brussels, 2002

20 The Lex column, *Financial Times*, 20th September 1999

21 'Review of the Charge Control on Calls to Mobiles'. Oftel, September 2001

22 'Use of Automatic Calling Equipment Review'. Statement issued by the Director General of Telecommunications, Oftel, January 2002

Chapter 10

Regulation and the future

10.1 Progress report

The development of the worldwide telecommunications services industry after nearly two decades of liberalisation, privatisation and regulation presents a mixed picture whose parts give rise both to satisfaction and disappointment. On the positive side, prices in this industry have plummeted both in nominal and real terms, while competition is now a reality in the long-distance, international and corporate markets of many countries. BT has less than 50 per cent of the long distance market for certain customer groups such as in the City of London, while at one stage MCI WorldCom was able to claim it had become a larger carrier than the former incumbent AT&T.

Nonetheless, incumbent dominance in the access sectors of most countries has barely been dented. BT is still dominant, retaining about 80 per cent of the UK residential access market, while many former monopolists have kept a greater portion than this of their access markets. BT still makes the bulk of its profits from traditional services. The bursting of the telecommunications investment bubble and the downturn in the telecommunications, media and technology (TMT) sector have seen potential competitors, once financially potent, now deeply indebted and desperately short of access to fresh investment funding. The market for long distance carriage has been confounded by spectacular improvements in the capability of dense wavelength division multiplex (DWDM) optical fibre technology. The resulting overhang of excess capacity, driven also by optimistic predictions of data volume growth, will exert a downward force on revenues for some years. This canvas presents a depressing picture to those who may have hoped that regulators would shortly be able to withdraw from effectively competitive markets.

10.2 Future regulatory strategy

10.2.1 The goal of 'light touch' regulation

Regulatory intervention in markets is not, of itself, a good thing, and should exist or continue to exist only if some other and greater evil would be incurred without it. This is true for two reasons, first because market competition is a better tool for promoting consumer welfare and economic efficiency, and second because regulation is both expensive and hazardous. Against this background, governments and the industry should have a general goal of 'light touch' regulation, that is, less regulatory activity. Wise regulators set themselves the same goal. While they will not, for reasons we shall see shortly, be advocating anything remotely like withdrawal, they will, therefore, be proceeding on bases of relevance, proportionality and appropriateness. As quoted in Chapter 3, the new UK regulatory body Ofcom may have an obligation not to impose or maintain unnecessary burdens and to periodically publish periodic statements showing how they propose to achieve this[1] [1]. The following list shows seven possible regulatory positions, in order of increasing 'lightness'.

- Government ownership and central control.
- Ongoing *ad hoc* sector-specific regulation.
- Sector specific regulation with ex-ante pre-emptive intervention where deemed appropriate under the principles of competition law.
- Sector-specific regulation based only on responsive ex-post action under the principles of competition law.
- Reliance on general competition law.
- Self-regulation within the industry.
- Withdrawal of regulation from the industry.

The majority of countries, certainly in Europe, migrated sometimes cautiously with market liberalisation in the 1980s and the 1990s from the first to the second of these. The development of European and national regulation since liberalisation has shown a trend towards the intellectual framework of competition law, positioning some of them near the third item in the list. Given the persistence of durable bottle-necks in this industry, that is of access networks, there remains an ongoing problem of dominance that will continue to require pre-emptive ex-ante regulation. The new European regulatory framework, coming into force on 25th July 2003, represents this third position in the list above.

10.2.2   The new European framework

The new European framework, described and paraphrased in some detail in the Appendix, provides a framework within which European Union countries will regulate their industries in accordance with unique national needs but in a harmonised manner. The framework is based on the principles of competition law, but assumes that there is a need for ongoing ex-ante regulation to deal with non-competitive markets. The new framework will bring changes in all the European Union countries, although the impact will be less disruptive in countries such as the UK that have already progressed some way from piecemeal and arbitrary intervention to the competition law basis.

At the heart of the framework are **market definition** [2] and **market analysis** [3] procedures for characterising the separate markets. As a harmonisation measure, the European Commission will recommend the definitions that should be adopted, and a fairly complex procedure will have to be followed for departure by a given member state regulator. Analysis classifies each market as competitive or non-competitive, as the case may be. In non-competitive markets, the player or players having market power must be identified.

Intervention in competitive markets is forbidden [4]. This is a major, new de-regulatory provision. Where a market is non-competitive, regulators may practise ex-ante regulation in accordance with stated policy objectives, which include *inter alia* consumer well being, universal services, encouraging efficient investment, supporting disabled users and an open European market [5]. Interventions themselves, that is remedies and obligations, are left for national regulators to decide from a permitted list. This includes transparency of tariffs, non-discrimination, accounting separation, mandatory granting of access to facilities, and finally price controls.

The Access Directive [6] deals with interconnection and access to facilities. National regulators are empowered to impose obligations on operators to grant access to specific facilities and services where denial would not be in the end users' best interest or would hinder the emergence of a sustainable competitive retail market [7]. Note that this obligation goes beyond the competition law concept of an essential facility that cannot readily be duplicated, to include one whose unavailability would hinder the development of competition. One hopes that the existence of the general test of consumer well being will prevent excessive intervention based on this article. Price control of such access is allowed where the supplying operator might be in a position to make excessive charges or apply a margin squeeze on competitors [8].

10.2.3 Price control paradigms

There appears to be a real opportunity for the withdrawal of regulatory intervention from retail price intervention by requiring vertically integrated operators with market power to provide the corresponding service on a wholesale basis. This opens the retail service to competitive supply, providing incentive regulation where the dominant player, its competitors and consumers all stand to benefit. So long as the wholesale price is determined on at least a transparent or a retail-minus basis, there is reduced scope for the dominant operator to apply a margin squeeze on the competitor. This is illustrated in Oftel's 2002 review of BT's retail price controls [9], applying a light retail price cap of RPI–RPI, that is constant nominal price, provided a wholesale line rental product is offered. The prospect of an even lighter cap is mooted, should competitors take up the wholesale product and thus generate real as opposed to theoretical competition.

Wholesale and interconnect markets remain all the more intractable while vertically integrated dominant players have control over essential infrastructure. Regulatory intervention as shown in Table 10.1 appears to be appropriate. Retail-minus price control prevents margin squeeze while allowing the dominant operator to make a worthwhile return on speculative investment, making it more likely that dominant players will be required to provide wholesale versions of innovative services. All this sadly implies plenty of ongoing regulatory activity, since it will entail a lot of work

Table 10.1   Wholesale price control options.

Wholesale market classification	Appropriate regulatory intervention
Non-competitive market, with dominance based on durable bottlenecks (e.g. access network)	Obligation to supply and cost-based price control
Prospectively or partly competitive market	Obligation to supply with retail-minus price control or non-discriminatory price requirement or Obligation to supply

to determine which categories apply in each case, and it will furthermore be far from straightforward to determine cost-based prices or to investigate for discrimination.

10.2.4   Vertical separation

Given the long-term persistence of the access bottleneck, it seems worthwhile to briefly revisit the issue of vertical separation. This is the division of a former incumbent monopolist into a 'Service Company' and a 'Network' or 'Loop Company', mirroring the structure that divides, for example, railway, gas and electricity industries into infrastructure and retail service entities. This possibility was advocated and no doubt considered at the time of privatisation of many publicly owned telecommunications operators. As explained in Chapter 4, most European countries eschewed it because of caution, to say nothing of lobbying by their incumbents, as it would have been fraught with problems.

It is difficult now to imagine vertical separation being seriously reconsidered. The same technical problems – that it is very difficult to draw the boundaries – still exist. To perform a technically fraught separation at a time when investment capital is scarce and innovative services are under development, would be extremely risky, to put it mildly. It is not clear how separation might be pursued under the new regulatory framework. It would certainly involve a great deal of government anti-trust process to make the case for enforcing it.

10.2.5   Operator bankruptcy

The problem of operator bankruptcy is hardly one that the industry regulators will have expected on the scale that has been seen since the downturn in the telecommunications sector. Of course, individual undertakings will always get into difficulties, but the market could be expected to correct over-supply and fragmentation through mergers, acquisitions, take-overs and consolidation. The issues that regulators face over large-scale operator bankruptcies are thus unexpected, and it may take some time for the best responses to be developed and understood.

A regulator has a duty to its consumers to ensure, as far as it is able, that they are not suddenly left without a supply of telecommunications service. In many cases, this will give it an incentive to want the troubled company to continue operating, or to transfer to new ownership as a going concern. On the other hand, the regulator needs to pay due attention to the market, which may be doing an efficient thing by driving that supplier from market and directing investment capital away from it. A company that shed some or all its debt after bankruptcy re-structuring might acquire a competitive advantage that threatened more efficient players, plunging the industry into potentially deeper woes.

Fresh dimensions enter the picture if a company stands accused of serious accounting fraud. Effective regulation depends on a degree of openness and integrity on the part of all the players in the industry. They will take part in discussions on how the industry should be regulated, and will enjoy an opportunity to express views on the price controls to be placed on the incumbent. These views are likely to be treated with respect, and may influence the regulation of the industry. Such a privileged position is incompatible with a propensity to fraudulent misrepresentation. The United States places a requirement of 'good character' on any holder of a broadcasting licence. While the position of a network operator is not directly comparable with the moral role of a content provider, there is still a valid question about fitness to operate a common carrier network, and this needs to be weighed with the other considerations.

10.3   New technology

10.3.1   Mobility

Technological innovation changes markets, sometimes disruptively, and an area showing plenty of active development is that of mobility. Mobile telephony has grown from a premium, professional market in the 1980s to a mass consumer market today. Many countries now have more mobile stations than fixed, wireline exchange connections. Some of this business is wholly new business, for example in the teenage and youth market, which before would have generated a much more modest number of calls from public telephone boxes. Other business may be substitutionary, especially in the market for residential second and third lines. The impact that third generation, 'mobile Internet', services may make is as yet unknown. Apart from the stimulation of wholly new markets, however, third generation may extend the challenge posed by mobile services to traditional wireline access networks. The question then for regulators is whether this may be the technological development that breaks the bottleneck of the ubiquitous access network. Any regulator who believes this to be a valid view will welcome the liberalisation of radio spectrum allocation as a key step along the road to a competitive telecommunications services industry.

10.3.2   Next generation networks

The advent of next generation network technology, briefly outlined in Chapter 9, is triggering a technical revolution in the way telecommunications networks are engineered. These changes will affect mobile as well as fixed local and long distance networks. Their impact will be to further reduce the whole life costs of networks, and also to bring about new service creation flexibility. This may simplify the separation of the transport and service functions in networks.

A challenge for regulators will be to ensure end-to-end service quality and many-to-many interconnection for the benefit of consumers. Regulators did not face major call quality concerns in the early days of market liberalisation because new entrants simply inherited through their equipment suppliers the technology that had been developed in monopoly for incumbents. A new generation of technology implies some degree of learning by experience, and a rising prominence for the subject of service quality. Where dominant players have an obligation to provide interconnection services or facilities to others on non-discriminatory terms, regulators will see their roles complicated by quality aspects of discrimination as well as the familiar financial factors. The time may not be far away when essential interface provisions have to be deployed to ensure inter-operability of next generation networks. Without this, larger operators able to control end-to-end connections may acquire a quality advantage giving them market power.

10.4   Note

1   Author's paraphrase.

10.5   References

1   UK Draft Communications Bill, paragraph 6 (House of Commons, London, 2002)
2   Directive 2002/21/EC of the European Parliament and of the Council of 7 March 2002 on a common regulatory framework for electronic communications networks and services (The Framework Directive), Article 15. The European Parliament, Brussels, 2002
3   Directive 2002/21/EC of the European Parliament and of the Council of 7 March 2002 on a common regulatory framework for electronic communications networks and services (The Framework Directive), Article 16
4   Directive 2002/21/EC of the European Parliament and of the Council of 7 March 2002 on a common regulatory framework for electronic communications networks and services (The Framework Directive), Article 16.3
5   Directive 2002/21/EC of the European Parliament and of the Council of 7 March 2002 on a common regulatory framework for electronic communications networks and services (The Framework Directive), Article 8
6   Directive 2002/19/EC of the European Parliament and of the Council of 7 March 2002 on access to, and interconnection of, electronic communications networks and associated facilities (The Access Directive). The European Parliament, Brussels, 2002

7   Directive 2002/19/EC of the European Parliament and of the Council of 7 March 2002 on access to, and interconnection of, electronic communications networks and associated facilities (The Access Directive), Article 12

8   Directive 2002/19/EC of the European Parliament and of the Council of 7 March 2002 on access to, and interconnection of, electronic communications networks and associated facilities (The Access Directive), Article 13

9   'Protecting consumers by promoting competition: Oftel's conclusions'. Oftel, June 2002

Appendix: the new European framework

A.1 Existing and historic directives

European Union directives on telecommunications regulation have the harmonisation of regulation among the member states as an objective. This is an additional purpose to that of regulation in general, and is aimed towards the creation of a single European market. European directives thus specify the type of regulation that should or should not be enacted in member states, and contain much in the way of administrative provisions. They do not reproduce precisely the content of national laws and regulatory rules, although they have a very large influence upon them.

The basic European approach to telecommunications regulation was established in 1990 with the following directives.

- Commission Directive 88/301/EEC of May 16, 1988 on competition in the markets in telecommunications terminal equipment.
- Council Directive 90/387/EEC of June 28, 1990 on the establishment of the internal market for telecommunications services through the implementation of open network provision.
- Commission Directive 90/388/EEC of June 28, 1990 on competition in the markets for telecommunications services.

These were subject to elaboration and development in various later directives, such as these.

- Council Directive 92/44/EEC of June 5, 1992 on the application of open network provision to leased lines.
- Commission Directive 94/46/EC of 13 October, 1994 amending Directive 88/301/EEC and Directive 90/388/EEC in particular with regard to satellite communications.
- Commission Directive 94/796/EC of November 18, 1994 on a common technical regulation for the pan-European integrated services digital network (ISDN) primary rate access.
- Council Directive 95/62/EEC of December 13, 1995 on the application of open network provision (ONP) to voice telephony.

- Commission Directive 96/2/EC of 16 January, 1996 amending Directive 90/388/EEC with regard to mobile and personal communications.
- Commission Directive 96/19/EC of 13 March, 1996 amending Directive 90/388/EEC with regard to the implementation of full competition in telecommunications markets.

The following directives of 1997 substantially revisited the regulatory framework.

- Directive 97/13/EC of the European Parliament and of the Council of 10 April, 1997 on a common framework for general authorisations and individual licences in the field of telecommunications services.
- Directive 97/33/EC of the European Parliament and of the Council of 30 June, 1997 on interconnection in telecommunications with regard to ensuring universal service and interoperability through application of the principles of Open Network Provision (ONP) (The 'Interconnection Directive').
- Directive 98/10/EC of the European Parliament and of the Council of 26 February, 1998 on the application of open network provision (ONP) to voice telephony and on universal service for telecommunications in a competitive environment (The 'Revised Voice Telephony Directive').

These in turn have been subject to ongoing development to take account of new developments.

- Directive 98/61/EC of the European Parliament and of the Council of 24 September, 1998 amending Directive 97/33/EC with regard to operator number portability and carrier pre-selection.

A further reappraisal of the telecommunications regulatory framework has led to a new set of four directives in 2002, which are described below.

A.2   The European regulatory framework

The European Parliament and Council of Ministers adopted on 7th March 2002 four new directives dealing with telecommunications regulation. These have to come into force, after transposition by national parliaments into member state law, by 25th July 2003. Repealing earlier directives, they consolidate much of the regulatory experience of the 1990s, though they break new ground most notably in two areas. These concern licensing, and the application of competition law principles to the definition of significant market power.

The licensing provisions provide for complete market liberalisation and will over time abolish licences altogether by allowing any number of players to enter a market under the terms of a general licence. Limitation by licensing of the number of players in a market will only be allowed when objectively necessary, for example in mobile markets dependent on the scarce resource of spectrum. Regulators have the obligation to analyse a market and designate certain players as having **significant market power**

(SMP). Once so designated, the company will then be subject to additional regulatory rules. Intervention in markets analysed as competitive will not be allowed.

The definition of significant market power was originally a structural measure, set at a 25 per cent market share of the market in question [1]. This served a simple purpose, to allow incumbent, former monopolists and some mobile operators to be designated and so subject to ex-ante regulation. More recent thinking has been suggestive of a revised figure of 40 per cent [2]. However, with markets becoming more competitive or prospectively competitive, the definition is shifting from simple structural measures towards the competition law concept of **dominance**, as amplified by European case law. This is defined (and elaborated by considerable discussion) in Reference 2 as follows.

> An undertaking shall be deemed to have significant market power if, either individually or jointly with others, it enjoys a position of economic strength affording it the power to behave to an appreciable extent independently of competitors, customers and ultimately consumers.

The new European directives, which will govern European telecommunications regulation in the coming years, are listed in Table A.1. Each directive commences with a **recital**, headed 'whereas . . .', stating the background and purposes of the directive. This background provides essential material for understanding the provisions and may

Table A.1   New (2003) European telecommunications regulatory directives.

Serial	Title	Outline purpose
2002/EC/19	Directive 2002/19/EC of the European Parliament and of the Council of 7 March 2002 on access to, and interconnection of, electronic communications networks and associated facilities (The Access Directive)	Access, interoperability and interconnection
2002/EC/20	Directive 2002/20/EC of the European Parliament and of the Council of 7 March 2002 on the authorisation of electronic communications networks and services (The Authorisation Directive)	Licensing and authorisation principles
2002/EC/21	Directive 2002/21/EC of the European Parliament and of the Council of 7 March 2002 on a common regulatory framework for electronic communications networks and services (The Framework Directive)	Methods of regulation
2002/EC/22	Directive 2002/22/EC of the European Parliament and of the Council of 7 March 2002 on universal service and users' rights relating to electronic communications networks and services (The Universal Service Directive)	Services and support of consumers

therefore assist in determining the performance or non-performance of obligations under the rules, should they come into question.

There now follows a summary of the provisions of each of the directives listed in Table A.1. This summary represents the author's paraphrase of the more important features. The original text, which can readily be obtained via the European Union's web site [3], is acknowledged and is in all cases complete, authoritative and definitive. Where editorial comments have been inserted in these summaries, these are distinguished by the use of square brackets.

A.3   The Framework Directive

Article 1: Scope and aims

The directive establishes a harmonised framework for the regulation of electronic communications services, electronic communications networks, associated facilities and associated services. It lays down tasks of national regulatory authorities and establishes a set of procedures to ensure the harmonised application of the regulatory framework throughout the Community.

Article 2: Definitions of terms
Article 3: National regulatory authorities

National regulatory authorities must be competent. They must be legally distinct and functionally independent of all the organisations providing services in the market. Where governments still control or own service-providing organisations, the regulatory function must be structurally separated from such ownership and control. Regulatory Authorities must operate transparently and impartially. Where the regulatory function is divided between different bodies, the division must be made publicly clear. The bodies must co-operate both with one another and with other bodies responsible for competition and consumer matters of common interest.

Article 4: Right of appeal

Member states must provide an effective, independent and competent appeal body. An organisation has a right of appeal to this body against any regulatory decision applying to it. The body must give reasons for its decisions, which must be subject to further judicial review if the body was not itself a court or tribunal.

Article 5: Provision of information

Organisations shall provide information as requested by the regulatory authority to enable it to do its job. The information requested must be proportionate to the purpose for which it is required, and the regulator must supply a reason for its being required. The regulatory body will respect commercial confidentiality, but apart from such confidentially should publish information that contributes to an open and competitive market.

Article 6: Consultation and transparency mechanism

Regulatory authorities must give interested parties the opportunity to comment on any instrument having a significant impact on the market. They must publish consultation procedures. They must provide a common point of public access for information on current consultations and the results of the procedure, subject to normal confidentiality.

Article 7: Consolidating the internal market for electronic communications

National regulatory authorities shall contribute to the development of the internal market [that is, within the European Union] by cooperating with each other and with the Commission. They must do this in a transparent manner to ensure the consistent application, in all member states, of the provisions of these directives. To this end, they shall seek to agree on the types of instruments and remedies best suited to address particular situations. Certain types of proposed measure [detailed in the article] must be referred to the Commission. If the Commission thinks the measure is a barrier to a single market or has doubts over its compatibility with Community law or the policy objectives in Article 8, it may require it to be withdrawn.

Article 8: Policy objectives and regulatory principles

[Reference is made to these important provisions from a number of places in the companion directives.]

Regulatory authorities shall pursue measures whose objectives will be the encouragement, *inter alia*, of the following policy objectives. They will do their utmost to make regulations technologically neutral.

- Maximum benefit for consumers in terms of choice, price, and quality. This is also to extend to special interest groups such as disabled users.
- Effective competition without restriction or distortion.
- Efficient investment in infrastructure and efficient use of scarce resources such as radio spectrum and numbering.
- An effective, open European market.
- Integrity and security of public communications networks.
- All citizens having access to a universal service [specified in the Universal Service Directive].
- Consumer protection in dealings with suppliers, including clarity and transparency in tariffs and conditions of use.
- A high level of protection of personal data and privacy.

Article 9: Management of radio frequencies for electronic communications services

Member states shall ensure the effective management of radio frequencies for electronic communication services in their territory. They shall ensure that allocation and assignment are based on objective, transparent, non-discriminatory and proportionate criteria. They may allow organisations with allocations to transfer them to other undertakings, subject to publication and to procedures to ensure that such transfer does not distort competition.

Article 10: Numbering, naming and addressing

Regulatory authorities shall control the assignment of national numbering resources and the management of national numbering plans. Adequate numbers and numbering ranges must be provided for all publicly available electronic communications services, with objective, transparent and non-discriminatory assignment procedures for national numbering resources.

Article 11: Rights of way

Competent planning authorities that have to consider applications for the granting of rights to install facilities on, over or under public and private property, will act on the basis of transparent and publicly available procedures. These will be applied without discrimination or without delay, and will follow the principles of transparency and non-discrimination when attaching conditions to any such rights. Procedures may vary depending on whether the applicant is providing public or private communications facilities. There shall be a right of appeal to an independent body.

Article 12: Co-location and facility sharing

Regulatory authorities shall encourage the sharing of facilities between organisations, and may impose it if necessary to protect the environment, or for public health, public security and planning objectives.

Article 13: Accounting separation and financial reports

Organisations providing public communications services that also have special or exclusive rights for the provision of other services must have structural separation or accounting separation. Their accounts must be as if the communications services had been carried out by independent organisations. This requirement may be waived for small organisations. Organisations that do not have to prepare accounts according to normal company law must prepare independently audited accounts.

Article 14: Undertakings with significant market power

[The principles to be used for determining whether an operator has significant market power or joint significant market power are laid down in this provision, including an annex and a reference.] An organisation having SMP in one market may be deemed to have SMP in a related market.

Article 15: Market definition procedure

The Commission will adopt and regularly review a definition of the different markets that may be held to exist within an overall national market for electronic communications services. National regulatory authorities may adopt market definitions that depart from this recommendation after following the requirements of Articles 6 (consultation and transparency) and 7 (market harmonisation).

Article 16: Market analysis procedure

Regulatory authorities must analyse the defined markets in accordance with EU guidelines. Where a market is competitive, it shall not impose specific regulatory requirements and must withdraw any currently in force. Where a market is not competitive, it will identify organisations with significant market power in accordance with Article 14, and impose or maintain specific regulatory obligations upon these organisations.

Article 17: Standardisation

The Commission will draw up a list of preferred standards and member states will encourage their use to the extent strictly necessary to ensure interoperability of services and to improve freedom of choice for users. The Commission has powers to request the following organisations to draw up standards:

- European Committee for Standardisation (CEN).
- European Committee for Electrotechnical Standardisation (CENELEC).
- European Telecommunications Standards Institute (ETSI).

In the absence of a recommended standard, member states will encourage use of European standards and in their absence international standards from these sources:

- International Telecommunication Union (ITU).
- International Organisation for Standardisation (ISO).
- International Electrotechnical Commission (IEC).

The Commission may make certain standards compulsory where necessary to ensure interoperability.

Article 18: Interoperability of digital interactive television services

To promote the free flow of information, media pluralism and cultural diversity, member states shall encourage providers of digital television and interactive television platforms and services to conform to an open Application Programming Interface (API).

Article 19: Harmonisation procedures

National regulatory authorities shall do their utmost to adopt Commission recommendations on the harmonised application of the provisions in the directive. Where they choose not to follow a recommendation, they must inform the Commission giving a reason. The Commission has powers to take implementing measures if this would result in a barrier to the single market.

Article 20: Dispute resolution between undertakings

If there is a dispute regarding the obligations of a provider of communications services, the national regulatory authority shall provide a binding resolution at the longest within four months. They may inform the parties if they intend some other method of resolution. Decisions will be made public (where confidentiality permits), the reasoning behind them must be stated, and the import of the decision must respect the principles of this directive (in particular Article 8) and the companion directives.

Article 21: Resolution of cross-border disputes

[This article is similar in intent to Article 20, buts adds that where the dispute lies within the competences of more than one national regulatory authority, then the authorities must coordinate their efforts.]

Article 22: Committee

A committee will assist the Commission. [Certain procedural aspects of the operation of the Committee are specified by reference to 1999/468/EC: 'Council Decision of 28 June 1999 laying down the procedures for the exercise of implementing powers conferred on the Commission'.]

Article 23: Exchange of information

The Commission shall provide all relevant information to the committee (above) on the outcome of regular consultations with network operators, service providers, users, consumers, manufacturers and trade unions, third countries and international organisations. The committee shall, taking account of the Community's electronic communications policy, foster the exchange of information between member states and the Commission on the development of regulation in electronic communications networks and services.

Article 24: Publication of information

Up-to-date information about the application of this and companion directives must be made publicly available with easy access for all interested parties.

Article 25: Review procedures

The Commission shall periodically review the functioning of this directive and report to the European Parliament and to the Council, on the first occasion not later than three years after the date of application (25th July 2003), and may request information from the member states.

Article 26: Repeal

[This article lists previous directives repealed by the application of this directive.]

Article 27: Transitional measures

All existing obligations laid on undertakings and all designations of significant market power will remain in force until the completion of the analysis procedures under Article 16.

Article 28: Transposition

Member states must enact laws, regulations and administrative provisions bringing the directive into effect by 23rd July 2002, and applying its provisions no later than 25th July 2003.

Article 29: Entry into force / Article 30: Addressees

[These are procedural articles stating that the directive comes into force when published, that it is enacted in Brussels for the European Parliament and Council, and that it is addressed to the member states.]

A.4   The Authorisation Directive

Article 1: Objective and scope

The directive applies to authorisations for the provision of electronic communications networks and services. Its aim is to implement an internal market through the harmonisation and simplification of authorisation rules and conditions.

Article 2: Definitions
Article 3: General authorisation of electronic communications networks and services

The provision of electronic communications networks or services may be subject only to a general authorisation. Member states may not prevent an undertaking from providing them except where necessary. The undertaking may be required to give notification, providing only so much information as is necessary for a regulatory authority to maintain a register of network and service providers. The undertaking may not be required to obtain a decision or administrative act before exercising its rights under the general authorisation.

Article 4: Minimum list of rights derived from the general authorisation

Authorised undertakings may:

- provide electronic communications networks and services;
- install facilities in accordance with the Framework Directive Article 11 (Section A.3);
- negotiate (or where applicable obtain) interconnection in accordance with the Access and Interconnection Directive.

Article 5: Rights of use for radio frequencies and numbers

Where there is no risk of interference, radio frequency usage shall not be subject to individual authorisation, although it may be subject to conditions of use. Where it is necessary to grant individual rights of use of radio frequencies, such rights of use shall be granted through open, transparent and non-discriminatory procedures. Decisions on rights of use shall be granted within three weeks for number ranges (this may be extended to six weeks when there are competitive selection procedures for numbers of especial economic value), and within six weeks for radio frequencies that have already been granted for specific purposes. Rights of use for radio frequencies shall not be limited except where necessary for efficient use of spectrum.

Article 6: Conditions attached to the general authorisation and to the rights of use for radio frequencies and for numbers, and specific obligations

A referenced annex lists the conditions that may be attached to general authorisations, and to rights of use of numbers and frequencies, whether general or individual. In the case of general authorisations, these include the following.

- Contributions to a universal service obligation.
- Administrative fees for the regulatory authority.
- Interoperability, interconnection and accessibility of numbers.
- Conditions linked to the granting of access to or use of public or private land, including environmental and town and country planning requirements.
- Data protection, consumer protection and illegal content requirements.
- Information to be notified.
- Technical standards requirements (Framework Directive Article 17, Section A.3).
- Electromagnetic radiation conditions.
- Terms of operation in times of major disaster.
- National security requirements.

In the cases of numbering and radio spectrum authorisations, permissible conditions include the following.

- Designation of the type of service or technology applying to the specific number or frequency range.
- Requirements and conditions necessary for effective and efficient use.
- Duration of authorisation.
- Conditions for transfer of rights.
- Usage fees.
- Any commitments made in the course of obtaining the rights under a comparative or competitive selection procedure.
- Obligations under international agreements.

Specifically in the case of frequencies, there may be technical conditions relating to interference. Specifically in the case of numbers, there may be conditions relating to portability and to provision of directory information.

Article 7: Procedure for limiting the number of rights of use to be granted for radio frequencies

This article lays down procedures that must be followed if there is to be a limitation of frequencies.

- The reasons for the limitation must be made public.
- Interested parties must be allowed to state a view.
- The limitation must be subject to periodic review.
- Allocation processes must be objective, transparent, non-discriminatory and proportionate.

Article 8: Harmonised assignment of radio frequencies

Where the usage of radio frequencies has been harmonised, member states must not add additional conditions, additional criteria or procedures that would restrict, alter or delay implementation of the common assignment of such radio frequencies.

Article 9: Declarations to facilitate the exercise of rights to install facilities and rights of interconnection

When requested by an undertaking, a national regulatory authority shall within one week issue a standardised declaration that a notification has been received from it, also confirming the circumstances under which it may construct facilities and negotiate (or obtain) interconnection.

Article 10: Compliance with the conditions of the general authorisation or of rights of use and with specific obligations

National regulatory authorities may require undertakings enjoying general authorisation and rights of use of numbers and spectrum to provide information necessary to verify compliance with authorisation conditions. In breach of a condition, the regulator must first notify the undertaking and require it to remedy the breach within a month of notification. If remedy is not forthcoming, the regulator may take proportionate enforcement measures including financial penalties. The reasons for the measure must be stated. In case of serious and repeated breaches, the regulatory authority may prevent the undertaking from continuing to provide service, and may withdraw its authorisations. Immediate action may be taken if public safety, public security or public health is threatened, or if the breach will create serious economic or operational problems for other providers or users. There must be a right of appeal (Framework Directive Article 4, Section A.3).

Article 11: Information required under the general authorisation, for rights of use and for the specific obligations

This article limits the rights of the national regulatory authority to require information:

- to that necessary for the regulator to perform its function;
- for clearly defined statistical purposes;
- to publish comparative price and quality reviews.

A reason must be stated for each requirement.

Article 12: Administrative charges

Regulatory authorities may impose an administrative charge, but this must be imposed upon individual undertakings in an objective, transparent and proportionate manner that minimises additional administrative costs and attendant charges. The total revenue is limited to that necessary to defray the management, control and enforcement of the general authorisation scheme and of rights of use and of specific obligations, and certain other items such as international co-operation.

Article 13: Fees for rights of use and rights to install facilities

Fees may be imposed for the rights of use for radio frequencies or numbers or rights to install facilities on, over or under public or private property. These should reflect the need to ensure the optimal use of these resources. Such fees shall be objectively justified, transparent, non-discriminatory and proportionate to their intended purpose and shall take into account the policy objectives in Article 8 of the Framework Directive (Section A.3).

Article 14: Amendment of rights and obligations

Rights, conditions and procedures concerning general authorisations and rights of use or rights to install facilities may be amended only in objectively justified cases and in a proportionate manner after the giving of no less than four weeks' notice. Rights may not be withdrawn before expiry except where objectively necessary, and then subject to compensation.

Article 15: Publication of information

Up-to-date relevant information on rights, conditions, procedures, charges, fees and decisions concerning general authorisations and rights of use must be made publicly available with easy access for all interested parties. Where information is held at different levels of government, the national regulatory authority shall make all reasonable efforts to create a user-friendly overview of all such information.

Article 16: Review procedures

[This is a similar provision to Article 25 of the Framework Directive (Section A.3).]

Article 17: Existing authorisations

Authorisations already in existence [for example, licences] shall be amended in accordance with this directive by the application date (25th July 2003). If this results in reduction or extension of existing obligations, then a nine-month extension is permitted. Where this would create excessive difficulties, for example where undertakings have benefited from mandated access to another network, member states may apply to the Commission for a temporary prolongation.

Article 18: Transposition

Member states must enact laws, regulations and administrative provisions bringing the directive into effect by 23rd July 2002, and applying its provisions no later than 25th July 2003.

Article 19: Entry into force / Article 20: Addressees

[These are procedural articles similar to Articles 29 and 30 of the Framework Directive (Section A.3).]

A.5   The Access and Interconnection Directive

Article 1: Scope and aim

The directive aims to establish a regulatory framework in accordance with internal market principles for the relationships between suppliers of networks and services that will result in sustainable competition, interoperability of electronic communications services and consumer benefits. It establishes rights and obligations for operators and for undertakings seeking interconnection with or access to networks or associated facilities. It sets out objectives for national regulatory authorities with regard to access and interconnection.

Article 2: Definitions
Article 3: General framework for access and interconnection

There shall be no restrictions preventing undertakings in the same or different member states from negotiating access and interconnection between themselves. An undertaking requesting access need not be authorised to operate in a member state if it does not operate a network or provide services there. There must be no legal or administrative measures imposing obligations that are not related to the access and interconnection services, or that require similar services to be offered on different terms to different other undertakings.

[Note that there is clear intent to liberalise international trade, preventing a country from entertaining different terms for undertakings of different nationality.]

Article 4: Rights and obligations for undertakings

Operators of public communications networks shall have a right and, when requested by other authorised undertakings an obligation, to negotiate interconnection for the purpose of providing publicly available electronic communications services. Information acquired from an undertaking during the course of negotiating access and interconnection must be kept confidential and used only for the purposes for which it was supplied. Digital television services and networks shall be capable of distributing wide-screen television.

Article 5: Powers and responsibilities of the national regulatory authorities with regard to access and interconnection

National regulatory authorities shall have power to require undertakings that control access to end users to interconnect. They may impose technical or operational conditions. These obligations and conditions shall be objective, transparent, proportionate and non-discriminatory. They must have the power to intervene on matters of access and interconnection at their own initiative and at the request of parties involved, in conformity with various articles in the Framework Directive (Section A.3). Obligations may be placed on operators of digital television and radio systems to provide access on fair, reasonable and non-discriminatory terms to Application Programming Interfaces (API) and Electronic Programme Guides (EPG).

Article 6: Conditional access systems and other facilities

Operators of conditional access services to digital radio and television are subject to a number of conditions as follows. These conditions may be reviewed from time to time, and may be withdrawn or amended if as a result of a market analysis one or more operators are found not to have significant market power.

- Systems must have cost-effective capability for control by operators at a regional or local level.
- Where there is dependence on the system to reach any group of potential viewers and listeners, access services must be offered to all broadcasters, on a fair, reasonable and non-discriminatory basis.
- Conditional access providers must keep separate accounts relating to their operations as conditional access providers.
- Holders of Intellectual Property Rights (IPR) relating to conditional access systems must, when granting licences to makers of consumer equipment, do so on fair, reasonable and non-discriminatory terms. They may not insert terms prohibiting, deterring or discouraging the inclusion in the same product of a common interface or access to another specific system.

Article 7: Review of former obligations for access and interconnection

Existing obligations to provide access and interconnection under earlier directives will remain in force until a market analysis, performed as soon as practicable in accordance with Article 16 of the Framework Directive (Section A.3), has determined that any should be maintained, amended or withdrawn.

Article 8: Imposition, amendment or withdrawal of obligations

National regulatory authorities must have the power to impose the obligations under Articles 9–13 of this directive. All obligations imposed shall be based on the nature of the problem identified, and be proportionate and justified in the light of the objectives laid down in Article 8 of the Framework Directive (Section A.3). They shall only be imposed following consultation in accordance with Articles 6 and 7 of that directive.

National regulatory authorities *will not* impose these obligations on undertakings that are not designated to have significant market power, though this is without prejudice to international commitments, personal privacy laws and other requirements in this and related directives. They *will* impose these obligations on undertakings designated as having significant market power. If they wish to add further obligations on undertakings with significant market power, they must submit a reasoned request to the Commission, which will authorise or prevent the proposed measures.

Article 9: Obligation of transparency

National regulatory authorities may require operators to make public specified information, such as accounting information, technical specifications, network characteristics, terms and conditions for supply and use, and prices. Where an operator has obligations of non-discrimination, national regulatory authorities may require that operator to publish a reference offer. They may prescribe the information to be contained within, and manner of publication of, a reference offer. The reference offer shall be sufficiently unbundled to ensure that undertakings are not required to pay for facilities that are not necessary for the service requested. It will give a description of the offerings broken down into components with the terms, conditions and prices. The national regulatory authority shall, *inter alia*, be able to impose changes to reference offers.

[A detailed annex to the directive specifies the minimum information to be provided in a reference offer for unbundled local loop services. Little of this is surprising. It has, however, the effect of *requiring* co-location and loop sharing.]

Article 10: Obligation of non-discrimination

A national regulatory authority may impose obligations of non-discrimination in relation to interconnection and/or access. These shall ensure, in particular, that the operator applies equivalent conditions in equivalent circumstances to other undertakings providing equivalent services. It will provide services and information to others under the same conditions and of the same quality as it provides for its own organisation, its subsidiaries and its partners.

Article 11: Obligation of accounting separation

A national regulatory authority may impose obligations for accounting separation in relation to interconnection and/or access. In particular, a national regulatory authority may require a vertically integrated company to make transparent its wholesale prices and its internal transfer prices. This is, *inter alia*, to allow the national regulatory authority to ensure compliance where there is a requirement for non-discrimination and, where necessary, to prevent unfair cross-subsidy. National regulatory authorities shall have the power to require that accounting records, including data on revenues received from third parties, are provided on request. They may publish such information as would contribute to an open and competitive market, subject to commercial confidentiality.

Article 12: Obligations of access to, and use of, specific network facilities

A national regulatory authority may impose obligations to grant access to and use of specific network elements and associated facilities. They may do this where denial of such (or the imposition of unreasonable terms and conditions having similar effect) would not be in the end users' best interest or would hinder the emergence of a sustainable competitive retail market. Requirements may include the following, taken together with conditions of fairness, reasonableness and timeliness.

- Interconnection of networks and network facilities.
- Third party access to specified network elements and facilities, including unbundled access to the local loop.
- Wholesale supply of specified services for resale by third parties.
- Open access to technical interfaces, protocols and virtual network services where essential for interoperability.
- Specified services necessary for end-to-end user services, including intelligent network services or mobile roaming.
- Access to operational support services and suchlike where necessary to ensure fair competition in service provision.
- Co-location and facility sharing, including sharing of ducts, buildings and masts.
- Good faith in negotiation.
- Non-withdrawal of facilities already granted.

Nonetheless, in assessing whether any obligation is proportionate to the policy objectives in Article 8 of the Framework Directive (Section A.3), national regulatory authorities may take into account the following things.

- Technical feasibility and economic viability, including in relation to the existing available capacity.
- Investments to be made by the facility owner, and the risks attaching thereto.
- The need to safeguard long-term competition.
- IPR where appropriate.
- Pan-European service provision.

Article 13: Price control and cost accounting obligations

A national regulatory authority may impose obligations relating to cost orientation of prices for the provision of interconnection and/or access. They may do this in situations where a market analysis indicates that a lack of competition might enable the operator concerned to sustain prices at an excessive level or apply a price squeeze, to the detriment of end users. However, national regulatory authorities shall allow the operator a reasonable rate of return on adequate capital employed, taking into account the risks involved.

Any cost recovery mechanism shall promote efficiency, sustainable competition and end user benefit. The burden of proof that charges are cost-oriented lies with the operator, who may be required to produce full justification. The regulator may require prices to be adjusted.

Article 14: Committee

[This is a similar provision to Article 22 of the Framework Directive, Section A.3.]

Article 15: Publication of, and access to, information

Up-to-date information pertaining to the obligations laid on operators under this directive, together with products, services and geographical markets, must be made publicly available with easy access to all interested parties. This does not apply, however, to confidential and commercially sensitive information.

Article 16: Notification

Member states must notify to the Commission the regulatory authorities responsible for the provisions of this directive by 25th July 2003. These must then notify to the Commission the undertakings deemed to have significant market power and the obligations laid upon them, and any subsequent changes thereto.

Article 17: Review procedures

[This is the same provision as Article 25 of the Framework Directive, Section A.3.]

Article 18: Transposition

Member states were to enact laws, regulations and administrative provisions bringing the directive into effect by 23rd July 2002, and applying its provisions no later than 25th July 2003.

Article 19: Entry into force / Article 20: Addressees

[These are procedural articles similar to Articles 29 and 30 of the Framework Directive (Section A.3).]

A.6   The Universal Service Directive

Article 1: Scope and aims

The directive aims to ensure the availability throughout the Community of good quality, publicly available services through effective competition and choice, and to deal with circumstances where the market might not satisfactorily meet the needs of end users. It establishes the rights of end users and the corresponding obligations on providers of publicly available electronic communications networks and services. It defines the minimum set of services of specified quality to which all end users shall have access at an affordable price.

Article 2: Definitions
Article 3: Availability of universal service

The services described in Articles 4, 5, 6 and 7 will be made available to all users at an (in the light of national circumstances) affordable price, regardless of geographical location. Member states must determine the most efficient and appropriate approach to their provision, in particular minimising the market distortion and safeguarding the public interest where the terms of provision of service depart from normal commercial conditions.

Article 4: Provision of access at a fixed location

All reasonable requests for fixed line and public telephone services must be met by at least one provider. The 'line' must have the following basic capabilities, taking into account the current state of technology and the facilities used by the majority of subscribers:

- local telephone calls;
- national and international telephone calls;
- fax communications;
- data communications at rates sufficient to support functional internet access.

Article 5: Directory enquiry services and directories

All users should have at least one directory available to them, updated at least annually, and a directory enquiry service also available at public telephones. These shall list all subscribers of public telephone services, and directory providers must apply non-discrimination in respect of information supplied them by other undertakings.

Article 6: Public pay telephones

National regulatory authorities can impose obligations on undertakings to ensure that public pay telephones are provided, in terms of geographical coverage, the number of telephones, quality of service and accessibility to disabled users. It may, however, not impose such an obligation if after consultation it determines that comparable services are widely available. All public telephones shall permit calls to the European emergency number 112 and other national emergency numbers without charge and without having to use any means of payment.

Article 7: Special measures for disabled users

Member states may take measures where appropriate to ensure access by disabled people to affordable telephone service, emergency access, directories and directory enquiry services equivalent to those enjoyed by other users. Bearing in mind national circumstances, they may take measures to ensure that disabled users have a choice of undertakings and service providers such as is available to the majority of users.

Article 8: Designation of undertakings

Member states may designate one or more undertakings to guarantee the provision of universal service. Undertakings or sets of undertakings may provide different service elements and in different geographical areas. The designation mechanism shall exclude no undertaking on principle, and shall be efficient, objective, transparent and non-discriminatory, so as to ensure cost-effective provision.

Article 9: Affordability of tariffs

Member states may impose price caps, common tariffs and geographical price averaging on universal services provided by designated undertakings under transparent and non-discriminatory conditions. They may require undertakings to support tariff options that depart from normal commercial conditions to support users on low income or with special social needs.

National regulatory authorities shall monitor the level of retail tariffs of the universal services identified in Articles 4, 5, 6 and 7 above, in particular in relation to national consumer prices and income.

Article 10: Control of expenditure

Designated undertakings must provide universal services under terms and conditions that do not require users to purchase services that are neither required nor necessary. Member states must allow their regulators to insist on provision of the following facilities to ensure that users can control their expenditure and avoid disconnection.

- Itemised billing.
- Selective outgoing call barring without a facility charge.
- Pre-payment terms.
- Phased payment of a connection charge.

Measures aimed to deal with non-payment should be published, proportionate and non-discriminatory and exercised with due notice. Apart from cases of fraud or persistent late payment or non-payment, service interruption should be limited to the service that is the subject of non-payment.

Article 11: Quality of service of designated undertakings

Designated undertakings shall specify and publish their quality of service for universal services in terms of parameters given in an annex to the directive. Regulators may set standards at least for basic fixed line services, and may specify additional standards relating to service provision for disabled users. Persistent failure of an undertaking to meet its standards may result in measures under the Authorisation Directive. [This refers mainly to enforcement powers under Article 10 of that directive.]

Article 12: Costing of universal service obligations

If a national regulatory authority is of the opinion that a universal service obligation may place an unfair burden on a designated undertaking, it may calculate the net cost bearing in mind any benefits that arise from being a provider of universal service. It may take into account any identified costs that came to light during the process (if any) for selecting the undertakings to be designated.

Article 13: Financing of universal service obligations

Where a national regulatory authority determines from its net cost calculation that there is an unfair burden on designated undertakings, the member state may finance the net costs, but no more, from public funds or by sharing the costs between providers of electronic communications networks and services. Any sharing mechanism must respect the principles of transparency, least market distortion, non-discrimination and proportionality, and be administered by the national regulatory authority or a body independent of the beneficiaries. Member states may, however, choose not to levy contributions from undertakings whose national turnover is less than a [stated] limit. Contributions must not be levied on any undertaking not providing service in that country.

Article 14: Transparency

Where a mechanism is in place for sharing the costs of universal service, its principles and details must be made public. National regulatory authorities should prepare an annual report detailing the costs borne and contributions made to universal service, subject to commercial confidentiality.

Article 15: Review of the scope of universal service

The Commission will carry out a review of universal service two years after the application date of the directive (25th July 2003) and thereafter at three-year intervals. This review shall examine social, economic and technological developments, taking into account, among other things, mobility and data rates in the light of the current technologies used by the majority of subscribers.

Article 16: Review of obligations

Existing obligations to provide access, universal service, carrier pre-selection and leased line provision will be maintained, pending a market analysis according to Article 16 of the Framework Directive (Section A.3). Any measures to amend or withdraw these obligations will require the authority of the Commission following the principles of Article 7 of the Framework Directive.

Article 17: Regulatory controls on retail services

When a market is analysed as not competitive and a national regulatory authority concludes that the measures of carrier selection and of the Access and Interconnection Directive (Section A.5) are insufficient to achieve the policy objectives of Article 8 of the Framework Directive (Section A.3), it may impose objective, proportionate and justified retail controls on undertakings designated as having significant market power. However, these measures must not be used when a market is deemed to be competitive.

The controls may include requirements and measures aimed at preventing:

- excessive pricing;
- predatory pricing;
- undue preference to specific end users;
- unreasonable service bundling.

The following types of price control may be employed.

- Retail price caps.
- Control of individual tariffs.

- Cost-oriented pricing.
- Pricing based on comparability with other markets.

Article 18: Regulatory controls on the minimum set of leased lines

Where a market for the provision of leased lines is not effectively competitive, under-takings with significant market power will be obliged to provide a minimum set of leased line services as listed by the Commission. The terms and conditions will follow the principles of non-discrimination, cost-orientation and transparency, and shall contain at least the information listed in an annex of the directive.

Article 19: Carrier selection and carrier pre-selection

Undertakings designated as having significant market power for the provision of fixed telephone service will enable their subscribers to access the public telephone services of any interconnected provider by carrier pre-selection, or by call-by-call selection based on a carrier selection code. Access and interconnection charging shall be cost-oriented, and any fee levied on subscribers must not be a disincentive to use the service.

Article 20: Contracts

This article lays down that consumers subscribing for connection or access to a public telecommunication network have a right to a contract. It lays down standards for the content and conduct of subscriber contracts.

Article 21: Transparency and publication of information

Transparent information about the prices, terms and conditions for access to and use of public telephone services must be made publicly available and kept up-to-date. An annex sets down the information to be published. Regulatory authorities shall encourage the publication of guides to enable consumers to evaluate the costs of alternative usage patterns.

Article 22: Quality of service

Undertakings providing publicly available services shall publish up-to-date, adequate and comparable information about their service quality. Regulators may specify the parameters, content, form and manner of information to be published.

Article 23: Integrity of the network

Member states must take all necessary steps to ensure the integrity of the public telephone network at fixed locations and, in the event of catastrophic breakdown or of *force majeure*, its availability. Undertakings must take all reasonable steps to ensure uninterrupted access to emergency services.

Article 24: Interoperability of consumer digital television equipment

Interoperability of digital television equipment is required. [An annex gives further details.]

Article 25: Operator assistance and directory enquiry services

All subscribers of public telephone services have the right to a directory entry, and to have access to a directory enquiry and operator assistance service. Undertakings which assign telephone numbers to subscribers shall meet all reasonable requests to make this information available for directory and directory enquiry purposes in an agreed format on terms which are fair, objective, cost-oriented and non-discriminatory. Member states may not restrict users in one state from accessing directory enquiries in another. All this is subject to personal data protection and privacy law.

Article 26: Single European emergency call number

All users of publicly available telephone services must be able to call the emergency number 112 free of charge. Calls to this number will be handled in the most appropriate way. Caller location information will be provided to the emergency authorities to the extent of technical feasibility.

Article 27: European telephone access codes

The standard international access prefix will be 00. This does not preclude existing or future special dialling arrangements for cross-border calls.

Article 28: Non-geographic numbers

End users in one state shall be able to access non-geographic numbers in another state where technically and economically feasible, except where the called subscriber has chosen to be accessible from a limited geographic area. ['Non-geographic' numbers include number translation services, personal numbers, special charge services and the like.]

Article 29: Provision of additional facilities

National regulatory authorities may insist on the provision of these facilities by all undertakings operating public telephone networks. These should be available between member states to the maximum feasible extent.

- Tone dialling, that is, Dual-Tone Multi-Frequency (DTMF) operation as specified in ETSI ETR 207.
- Calling Line Identity (CLI), though subject to protection of personal data and privacy.

Article 30: Number portability

All subscribers of publicly available telephone services including mobile services shall be able if they request to retain their telephone number independently of the providing undertaking. This is at a fixed location in the case of geographic numbers or at any location in other cases. Interconnection charges relating to portability shall be cost-related. Subscriber fees must not act as a disincentive to taking portability. Regulators must not impose fixed charges that distort competition, for example a fixed retail tariff.

Article 31: 'Must carry' obligations

Member states may impose obligations for the transport of specified television and radio channels on networks where significant numbers of people depend on them as their main means of receiving television and radio.

Article 32: Additional mandatory services

Member states may make additional services publicly available beyond the universal services defined in Articles 4–7 above but may not impose a compensation mechanism for them.

Article 33: Consultation with interested parties

National regulatory authorities must take account of the views of end users, and consumers (including disabled users), manufacturers, providers of electronic communications networks and services, on issues related to end user and consumer rights. Interested parties may develop with the guidance of national regulatory authorities mechanisms to improve the general quality of service by, for example, setting and scrutinising codes of practice and operating standards.

Article 34: Out-of-court dispute resolution

Member states shall make available transparent, simple and inexpensive out-of-court procedures for dealing with unresolved disputes involving consumers.

Article 35: Technical adjustment

The Commission will handle adjustments necessary to adapt annexes of the directive for technological change. [References given in Article 37 provide further detail about the procedure.]

Article 36: Notification, monitoring and review procedures

The Commission shall periodically review the functioning of this directive and report to the European Parliament and to the Council, on the first occasion not later than three years after the date of application (25th July 2003), and may request information from the member states.

National regulatory authorities shall notify to the Commission the undertakings designated as having universal service obligations by 25th July 2003 together with the nature of these obligations, and any changes thereto as soon as possible. They shall similarly notify undertakings deemed to have significant market power for the purposes of this directive, and their obligations.

Article 37: Committee

[This is a similar provision to Article 22 of the Framework Directive (Section A.3).]

Article 38: Transposition

Member states must enact laws, regulations and administrative provisions bringing the directive into effect by 23rd July 2002, and applying its provisions no later than 25th July 2003.

Article 39: Entry into force / Article 40: Addressees

[These are procedural articles similar to Articles 29 and 30 of the Framework Directive (Section A.3).]

A.7   References

1   Directive 97/33/EC of the European Parliament and of the Council of 30 June 1997 on interconnection in telecommunications with regard to ensuring universal service and interoperability through application of the principles of Open Network Provision (ONP)
2   'Draft guidelines on market analysis and the definition of significant market power'. EC Working Document COM (2001) 175, March 2001. The European Commission, Brussels
3   European legislation may be obtained at *http://europa.eu.int/eur-lex*

Index